OCCUPATIONAL SAFETY
AND HEALTH SERIES
No. 42

BUILDING WORK

A COMPENDIUM OF OCCUPATIONAL SAFETY
AND HEALTH PRACTICE

INTERNATIONAL LABOUR OFFICE   GENEVA

ISBN 92-2-101907-1

First published 1979

The designations employed in ILO publications, which are in conformity with United Nations practice, and the presentation of material therein do not imply the expression of any opinion whatsoever on the part of the International Labour Office concerning the legal status of any country or territory or of its authorities, or concerning the delimitation of its frontiers.
The responsibility for opinions expressed in signed articles, studies and other contributions rests solely with their authors, and publication does not constitute an endorsement by the International Labour Office of the opinions expressed in them.

ILO publications can be obtained through major booksellers or ILO local offices in many countries, or direct from ILO Publications, International Labour Office, CH-1211 Geneva 22, Switzerland. A catalogue or list of new publications will be sent free of charge from the above address.

Printed by the International Labour Office, Geneva, Switzerland

FOREWORD

Building comes high in the comparative table of accident rates in the different industries, a close third after the stone and earth industry and mining. This relatively high accident frequency is not caused entirely by the fact that work has often to be done at great heights or in deep pits or shafts: the very nature of building operations with their constantly changing conditions makes a specific risk structure.

Following the recommendation of the Building, Civil Engineering and Public Works Committee of the International Labour Organisation at its Seventh Session (Geneva, May 1964) expressed in a Resolution (No. 69) the Office prepared a Code of Practice which is entitled "Safety and Health in Building and Civil Engineering Work". This Code of Practice was approved for publication by the Governing Body of the ILO at its 180th Session (May-June 1970).

In accordance with its usual practice of publishing explanatory material to illustrate and amplify the material contained in codes of practice, the Office prepared the present publication with the assistance of its Panel of Consultants on Occupational Safety and Health in Building, Civil Engineering and Public Works. In view of the wide range of subjects to be dealt with, it was decided to split the subject into two parts, of which the present work relates to the building industry. The ILO is indebted to Mr. A. von Chossy (Federal Republic of Germany) for having prepared the draft manuscript of this publication. A companion publication, covering safety and health in civil engineering work will be published.

CONTENTS

	Page
INTRODUCTION: Accident risks in the building industry	1

TECHNICAL ACCIDENT-PREVENTION MEASURES

1. Operations 5
   - 1.1 Building operations generally 5
   - 1.2 Excavations 17
   - 1.3 Underpinning and shoring operations 27
   - 1.4 Work with concrete and reinforced concrete 30
   - 1.5 Erection of prefabricated parts 35
   - 1.6 Work at and on roofs 38
   - 1.7 Work on chimneys 40
   - 1.8 Demolition 44
   - 1.9 Other operations 47

2. Equipment 65
   - 2.1 Machinery 65
   - 2.2 Electrical installations 80
   - 2.3 Lifting appliances, building hoists, lifting gear . 87
   - 2.4 Mechanical conveyance 118
   - 2.5 Scaffolds 123
   - 2.6 Working platforms 153
   - 2.7 Ladders, ladderways, stairs, gangways 156
   - 2.8 Tools 163
   - 2.9 Other industrial equipment 169

3. Dangerous gases and material 186
   - 3.1 General 186

			Page
	3.2	Health risks	186
	3.3	Fire and explosion risks	190
4.	Personal protective equipment		194
	4.1	General	194
	4.2	Hard hats	194
	4.3	Eye protection	195
	4.4	Safety boots	199
	4.5	Protective gloves	201
	4.6	Respiratory protective equipment	201
5.	Safety harness - catch nets		205
	5.1	Safety harness	205
	5.2	Catch nets	209

HEALTH PROTECTION

			Page
6.	Diseases due to lead and its compounds		211
	6.1	General	211
	6.2	Protective measures	211
	6.3	Medical supervision	212
7.	Carbon monoxide poisoning		213
8.	Disease due to dust		214
	8.1	General	214
	8.2	Course of the disease	214
	8.3	Protective measures	214
9.	Skin diseases		218
	9.1	General	218
	9.2	Prevention of occupational skin diseases	218

			Page
	9.3	Cleansing the skin	220
	9.4	Procedure in cases of skin disease	220
10.	Protection against noise		221
	10.1	General	221
	10.2	Noise levels	221
	10.3	Means of reducing noise at building sites	222
11.	Diseases due to vibration in work with compressed air		225
	11.1	General	225
	11.2	Prevention of diseases due to vibration in work with pneumatic tools	225

HYGIENE, FIRST AID, WELFARE

12.	Hygiene		227
	12.1	Workers' accommodation	227
	12.2	Washing facilities	228
	12.3	Toilets	228
	12.4	Other facilities	228
	12.5	Transportable accommodation	229
13.	Measures of occupational hygiene		230
14.	First aid in occupational accidents		231
	14.1	General	231
	14.2	Instruction in first aid in accidents	231
	14.3	First-aid supplies, rescue apparatus	237
	14.4	First-aid personnel	238
15.	Medical care		239
	15.1	General	239

		Page
15.2	Works medical service	239
15.3	Qualifications and functions of the works doctor	239

THE ORGANISATION OF WORK

16.	General		241
17.	Organisation of undertakings and building operations		242
	17.1	General principles governing the organisation of undertakings	242
	17.2	Planning and preparation of work	242
	17.3	Building site equipment	243
18.	Employment of personnel		246
	18.1	General	246
	18.2	Qualifications of workers	246
	18.3	Training and retraining	248
	18.4	Duties of workers	248
19.	Management and supervision		251
	19.1	Appointment of responsible managers	251
	19.2	Establishment of supervision	251
20.	Collaboration of the works council		252
21.	Organisation of safety		253
	21.1	General	253
	21.2	Works safety organisations and their tasks	253

INTRODUCTION

Accident risks in the building industry

Prevention of accidents and occupational diseases requires an exact knowledge of the sequence of causes, for only then can the necessary preventive measures be correctly chosen and fully applied. Owing to the constantly changing conditions as well as the great variety of equipment and processes in building operations the nature of accidents varies much more than might be supposed.

Building accidents may occur in any of the following ways:

1. Through the collapse of walls, building parts, stacks, masses of earth.

2. Through the collapse and overturning of ladders, scaffolds, stairs, beams.

3. By falls of objects, tools, pieces of work.

4. By falls of persons from ladders, stairs, roofs, scaffolds, buildings; through hatches and windows; into openings; on the level.

5. During loading, unloading, lifting, carrying and transporting loads.

6. On or in connection with vehicles of all kinds.

7. In the operation of railways.

8. At power plant and power transmission machinery.

9. At working machines.

10. On lifting and transport appliances.

11. On welding and cutting equipment.

12. On compressed-air equipment.

13. In connection with combustible, hot or corrosive materials.

14. In connection with dangerous gases.

15. During blasting with explosives.

16. When using or handling hand tools.

17. In connection with traffic on the building site.

18. On the way to and from work.

Among occupational diseases of building workers should be mentioned:

1. silicosis in stonesmasons, furnace lines and sand blasters;

2. lead poisoning in painters;

3. diseases of the joints and bones in workers operating pneumatic tools;

4. poisoning by carbon monoxide and benzene;

5. skin diseases which necessitate a change of occupation (e.g. in cement workers).

There are as many possible causes of accident as there are occasions. Among these are technical defects in equipment and methods of work, defects in organisation, dangerous acts by workers, that is the same defects that may be found in any undertaking. To these have to be added those causes that come from the nature of building operations themselves, defects in planning and construction or associated with constant change of the workplace or the friction usually found when different trades are working in close proximity to each other.

A classified list of these accident causes might be as follows:

I. Planning, organisation

(a) defects in technical planning;

(b) fixing unsuitable time limits;

(c) assignment of work to incompetent contractors;

(d) insufficient or defective supervision of the work;

(e) lack of co-operation between different crafts.

II. Execution of the work

(a) constructional defects;

(b) use of unsuitable building materials;

(c) defective processing of building materials;

(d) other defects.

III. Equipment

(a) lack of necessary equipment;

(b) use of unsuitable equipment;

(c) constructional defects in equipment;

(d) lack of safety devices or measures.

IV. **Management and conduct of work**

(a) inadequate preparation of work;

(b) inadequate examination of equipment;

(c) imprecise or inadequate instructions concerning work;

(d) employment of unskilled or insufficiently trained workers;

(e) inadequate supervision of work.

V. **Workers' behaviour**

(a) irresponsibility;

(b) unauthorised actions;

(c) carelessness.

Accidents on the work site have many causes and the preventive measures should match these: it is not enough to eliminate only one cause. Many measures may be necessary as one accident may have several causes: there should be a detailed risk analysis for every job. Where there are safety rules or regulations they will give a basis for this analysis but when new equipment and processes for which no such regulations exist are concerned, the risks should be analysed in each individual case so that safe procedures may be established.

A distinction may be made between technical measures and those concerning organisation, education and training. In the technical measures, safety devices play an important part and so do the various specifications and standards for materials, appliances and structures. Job instructions and training in safe methods of work, reinforced by competent supervision, are also essential.

Organisation for accident prevention should start in the planning of an undertaking and extend to all its activities. It should include the appointment of competent supervisory staff and the arrangement of training. All safety measures should also be the concern of the undertaking's safety organisation.

TECHNICAL ACCIDENT-PREVENTION MEASURES

1. Operations

There is a distinction between technical means of securing the safe condition of equipment and safe methods of work. The most important technical measures are dealt with in the sections on equipment and operations but these terms do not define completely separate fields: regulations concerning equipment may also cover its operation and use and regulations concerning operations may also deal with the design and fittings of some types of equipment.

1.1 Building operations generally

"Building operations" cover all those operations not carried on in a permanent workplace but at a building site. Operations carried on in the danger zone of a building site require accident-prevention and health-protection measures specially adapted to the nature of the building industry.

General

To meet the special needs of the building industry it is necessary to bear in mind how the nature of the industry affects safety. Risks in the building industry are higher than in industries with fixed workplaces for a number of reasons:

- usually the work has to be done in the open air and in all sorts of weather - heat, cold, rain, snow;

- defects in planning and execution and the use of unsuitable building materials jeopardise the stability of the building or parts of it;

- repeated erection and dismantling and changes in use lead to greater wear and tear and increase the risk;

- semi-skilled or quite unskilled workers are employed to a greater extent than in industries with fixed workplaces;

- supervision is considerably hampered by the size of the site and the difficulty of keeping the workplaces in sight;

- owing to the variety of the jobs, even specialists and supervisors have sometimes to take decisions that are beyond their powers because they do not possess the necessary special experience;

- some equipment and some protective measures are needed for only a short time as the work progresses and are makeshift in character and in breach of regulations.

These factors, singly or together, may lower the level of safety on the site very considerably. It is important that all measures for accident prevention and health protection should be carried out scrupulously.

Requirements for workplaces and passageways

The word "workplace" may have a variety of meanings at a building site - the floor of a trench or a foundation pit that has been excavated, the scaffold on which work has to be done, a ladder, a building part, a roof, the operator's stand on a machine or a lifting appliance or any other place from which work has to be done. Much the same applies to passageways on building sites: these include not only paths at ground level, but also ladderways, stairs, gangways and upper floors used for access to workplaces. These are things that must not be overlooked when the safety of workplaces or passageways is in question. Workplaces and passageways should always be kept in good condition. They should never be makeshift even if they are needed for a very short time. They should be kept clear of tools, materials and other obstructions, and protected against the risks of falling objects and persons. They should be lighted as conditions at the place and time require. Workplaces that are continuously occupied, for example, operators' stands on machines, should have means of heating in cold weather.

If owing to the nature of the work or the ground, or the weather conditions, workplaces and passageways cannot be prevented from becoming slippery suitable measures against slipping should be taken. When the season or the weather so requires appropriate measures should be taken to prevent draughts.

In the vicinity of dangerous places such as moving machine parts, drives, rail tracks, electric conductors, dangerous building parts, slopes and places under scaffolds, work should only be done, and passageways laid out and used, after the necessary precautions have been taken.

Protection against falls and throwing down of objects

At all places where there is a risk of building materials, tools, etc. falling and injuring people, sufficiently wide and strong protective roofs should be installed. They should be strong enough to prevent objects falling through them. Where such precautions cannot be taken people should be prevented from staying or passing under dangerous places by means of barriers. This also applies when horizontal nets used as protection against falls of persons have a mesh wide enough to allow objects to fall through. Tools should not be placed, even temporarily, in places where vibrations or unexpected movements might cause them to fall. Objects should only be thrown down if the danger zone has been guarded or otherwise reliably protected. No object should be thrown before a loud warning signal has been given and the thrower has satisfied himself that the danger zone is clear and barred. Guards who have been appointed to watch the danger zone and its vicinity should have no other duties and not concern themselves with anything else. Since lives may depend on the reliability of these guards they must be carefully selected and properly instructed.

1.1.1 Protective hoardings

Protective hoardings installed at the edge of a sloping roof should project at least 1 m beyond the edge, and be capable of safety retaining a weight of 150 kg rolling off a roof 5 m high sloping at 60º.

1.1.2 Catch nets

See section 5.2 on safety harness and catch nets.

1.1.3 Safety belts

See section 5.1 on safety belts.

Other requirements

1.1.4 Fragile building parts

Building parts that cannot support weights should not be loaded nor should weights be thrown or persons jump on surfaces that cannot safely support them.

1.1.5 Machine protection

In the operations and servicing of machines the relevant instructions should be complied with. See section 2.1 on machines.

Machines and other appliances should be provided with the necessary protection even if they remain idle for long periods. This requirement only ceases to apply when a machine is incapable of working.

Unauthorised persons should not interfere with machines. If a worker has to use another firm's machines, which often happens in the building industry where a number of firms work side by side, the permission of the employer concerned or his authorised representative should first be obtained. It should also be verified in advance that the worker is familiar with the operation of the machine.

1.1.6 Clothes and hair

Clothes and hair should be so arranged that they do not endanger the wearers. Clothes should be comfortable and suited to the weather conditions. People should not go about barefoot on building sites. See also section 4.4 on foot protection.

Clothes should not be changed or kept at dangerous places.

1.1.7 Smoking and drinking

Persons should not smoke at places where there is a risk of fire, explosion or the like.

Wholesome drinking water should be provided at all building sites, with suitable drinking vessels. No alcoholic drinks should be provided at building sites or in workshops.

Precautions against falls of persons

1.1.8 General considerations and requirements

Falls of persons are mong the most typical accidents at building sites. These accidents include both falls from heights, that is from buildings, scaffolds, ladders, etc., and falls through openings in floors, working platforms or scaffolds, or into excavations and shafts. Consequently exceptional importance attaches to precautions against falls of persons irrespective of the height of a possible fall. A considerable number of persons are killed by falling from only small heights. Precautions against falls of persons should apply from the lowest height at which it seems reasonable, which is usually taken to be 2 m.

Among precautions against falls of persons, a distinction can be drawn between collective precautions and individual precautions, that is between precautions that consist of arrangements designed to afford quite general protection, and precautions that an individual has to take on a certain job for his own protection. Collective precautions include railings with toeboards and retaining devices in the forms of horizontal protective scaffolds, vertical and inclined hoardings, and catch nets. An individual precaution is wearing a safety belt with a life line.

The following general requirements apply to precautions against falls of persons:

- safety devices should be so designed that they satisfy the requirements applying to them, and provide really effective protection;

- the installation, use and removal of a safety device should not involve a greater danger than the one it was designed to avoid;

- in deciding the precautions to be taken in any particular case, collective precautions should always be preferred to personal protection. For example, a safety belt with a life line should only be used when devices such as railings, hoardings or nets cannot be used, or their use would be too costly on the particular job to be done;

- if one precaution has to be withdrawn, even temporarily, because it hampers the work it should be replaced by another. Persons should never be left without any protection.

1.1.9 Precautions for different kinds of work

The suitability of precautions taken against falls of persons will depend on the nature of the work to be done and conditions on the spot. The following requirements apply in the circumstances described below.

(a) **Work on scaffolds, working platforms and working floors**

See also section 2.5.

If they are more than 2 m above the ground or any other sufficiently broad surface, scaffolds, working platforms and working floors should always be fenced on all open sides with railings and toeboards. This does not rule out additional precautions if there is a risk of materials or tools falling (Fig. 1).

If the space between the scaffold or the working floor and the building exceeds 30 cm, railings and toeboards must also be installed on the side facing the building, for otherwise the persons employed on the scaffold or platform would be in danger of falling through this space. There is also a risk of falls if there are openings in the building walls, and floor inside the building is more than 2 m below the scaffold floor at the level of the scaffold or the working platform. In such cases also railings and toeboards should be installed on the inner side of the scaffold or platform.

(b) **Work on floors or on scaffolds erected on floors**

There is a risk of falling outwards when work is being done from the floors of a building or from scaffolds erected on these floors. Consequently, for all work that requires a person to remain on the edge of a floor even for a short time safety devices should be provided. These devices may be horizontal, vertical or inclined retaining surfaces which should not be more than 2 m below the workplaces. If such outside protective and retaining devices cannot be installed as when prefabricated reinforced concrete units are being erected, or when the slab type of construction is adopted, railings and toeboards should be fitted at the edges of the floors, provided that they can be securely fastened. It may be necessary to remove these fastenings easily and this may require special measures to be taken.

(c) **Work at and on roofs**

See also section 1.6.

So far as falls of persons are concerned work at or on roofs is considered to be particularly dangerous. If outside scaffolds are used, care must be taken that they are provided with the necessary safeguards for roof work, which means that they must be specially designed for such work. In particular, they should be provided with a vertical hoarding or protective netting whose top edge is at least 1 m above the edge of the roof. The top floor of the scaffold should not be more than 1.5 m below the roof edge.

In the absence of outside scaffolds, horizontal, vertical or inclined retaining hoardings or platforms should be erected so that persons cannot fall from the roof over their edge. If suitably arranged, catch nets can serve the same purpose. If roof work is done from roof ladders or roofers' scaffolds these should be so fastened that they cannot slip down or sideways, and retaining devices at the edge of the roof will still be required.

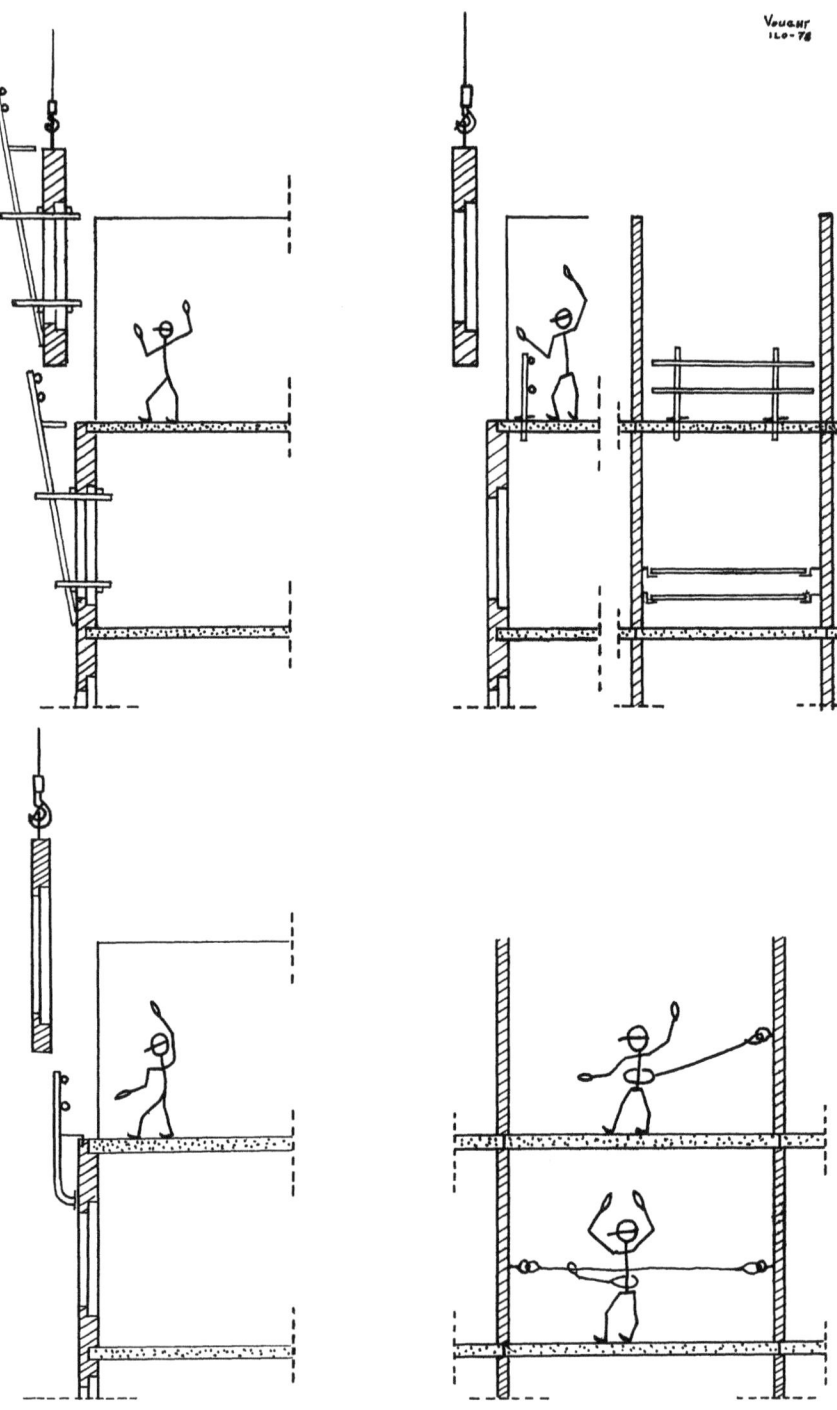

Fig. 1
Fencing of floor edges

If no work has to be done on the edge of the roof vertical hoardings 1 m high can be hooked on to the edge. For roof maintenance work of short duration it will be sufficient to secure the workers with a safety belt and a life line, but the point for attaching the line should be examined beforehand to make sure that it is safe. This is specially important when permanently fixed roof hooks for subsequent maintenance work are used for attaching roof ladders or life lines.

When work is done on fragile roofs, such as asbestos-cement roofs and glass roofs, in addition to precautions against falls over the edge of the roof, precautions should be taken against falls through it. Care must also be taken that the work can be done from special equipment so that the fragile roof does not have to be stepped on, and indeed cannot be stepped on even accidentally. One precaution is to instal gangways on the edge of the roof and not over it. Another is to place a catch scaffold, wire grating or a net immediately under the roof in case anyone inadvertently steps on it.

The precautions with work on flat roofs are the same as those described for work on floors.

Among other measures for preventing falls during work on roofs may be mentioned the wearing of suitable footwear, such as shoes with rope soles.

(d) Erection of prefabricated parts including roof trusses

See also section 1.5.

The risk of falls of persons is higher than average among building operations in the erection of structures such as prefabricated parts and roof trusses. The nature of these operations makes it extraordinarily difficult to take precautions but this does not mean that they can be dispensed with. On the contrary, these operations require careful consideration of the manner in which the demands of accident prevention can be met in each particular case. This consideration should be given in the planning stage, when the form and weight of the prefabricated parts and also the method of erection are determined. When deciding the dimensions of parts, arrangements for transporting them should not be overlooked, and they should be provided with means of attaching them to lifting appliances and of affixing safety devices. The erection procedure should be settled in all its details at the planning stage including the precautions to be taken in each phase. These precautions should be set out in special erection instructions for the site.

If catch nets or catch scaffolds are installed under the workplace it should be ensured that they are moved up as the work proceeds: the use of mobile nets and scaffolds makes this easier. If the building parts have to be stepped on during erection or if they serve as a workplace, adequate protection can be afforded by railings, provided that due preparations have been made for affixing them and they are put in place before the parts are transported to the building site.

For work of short duration the workers may wear safety belts with life lines, provided that there are suitable anchorages for the lines. Life lines should be kept as short as possible so that if a person falls he will not be in danger of striking against a part of the building. Often it will be useful to fasten the snap hook of the belt directly to an anchorage prepared on a part of the building.

When railings, catch scaffolds, catch nets and safety belts cannot be used the work can be done from a boatswain's chair raised to the workplace by a lifting appliance. Both the lifting appliance and the boatswain's chair should comply with the relevant constructional regulations.

(e) Demolition operations

See also section 1.8.

The precautions that may be taken against falls of persons in demolition operations are as various as the methods of demolition. Where buildings or walls are to be pulled down the greatest risks threaten the man who has to put the wire rope round the part to be pulled down. On this work, which is usually of short duration, a mobile mechanical ladder can usefully be employed. It is often not possible to use single portable ladders instead because of the instability of the parts to be pulled down.

If buildings are to be broken down the work should be done from work scaffolds. Workers should stand on walls only when they are thick enough to afford a good foothold and the drop is not more than 2 m.

When assemblages have to be taken to pieces the work should be done in the same way as described for the erection of prefabricated parts.

(f) Erection of scaffolds

See also section 2.5.

Precautions in scaffold erection usually cover both precautions to ensure the stability of the scaffold itself, and precautions to ensure the safety of the workers employed on it.

There are, however, risks of falls in the erection and dismantling of scaffolds. One of the most important safety rules for these operations is that the prescribed safeguards for work on scaffolds (railings, toeboards, close-planked floors) should be installed early enough for them to afford protection against falls of the workers erecting a scaffold. On high scaffolds protection against falls over the side may also be provided by installing inclined catch platforms or catch nets at intervals of from 8 to 10 m. It is equally important that when a scaffold is being dismantled the safeguards and the floors should not be removed until they are no longer needed to protect the dismantling crew.

Workers erecting or dismantling extensive scaffolds should be equipped with safety belts and life lines that will enable them to fasten themselves to parts of the scaffold when they have to work at particularly exposed places even for only a short time.

(g) **Work from ladders**

See also section 2.7.

If work has to be done at a height, and if the work conditions on the spot allow, a workplace in the shape of a work scaffold or a working platform should be provided. Work should only be done from ladders when a scaffold or a platform cannot be provided. Ladders should be placed firmly in position and secured against slipping, turning over and excessive sagging. The nature of the protection against slipping will depend on the condition of the ground. To prevent excessive sagging the uprights should be braced. The ladder should be prevented from slipping sideways by tying it or other means.

Double step ladders should not be more than 5 m high. They should be secured against overturning and sliding apart of the sides. To provide a better foothold bracket-like steps can be fastened to the uprights. To enable the user to hold on to the ladder with one hand he should carry his tools and other equipment in a container that can be hung on to the ladder.

Persons working on travelling mechanical ladders should be roped if there is no fenced working platform.

(h) **Other matters**

Work at the unloading points of lifting appliances

See also section 2.3.

Of the many risks attending the operation of lifting applicances not the least is that of falling from an elevated unloading point. Falls from such points can be caused in various ways. A person employed there may slip, or stumble or be struck by a swinging load. He may also be endangered when a hook is prematurely lowered while he is detaching or pulling in the load, or when he leans outward to pull in the load. If for certain types of lifting appliance, such as building hoists, the relevant regulations do not require more comprehensive precautions against falls, such as self-raising doors, mechanical interlocks, and blocking devices, persons at the unloading point should be protected by railings and toeboards. Where railings have to be removed for short periods for landing the load, and a mechanical device would not be effective, care should be taken that it is replaced immediately after the load has landed. To avoid the need for leaning outwards a hook should be provided with which the load can be pulled on to the landing. Where collective precautions cannot be taken persons employed at the unloading point should be roped.

Work near excavations and shafts, and openings in floors and platforms

All openings in floors and working platforms, and also edges of excavations and shafts in working and traffic areas should be protected by railings and toeboards irrespective of the height of a possible fall, unless they are provided with a sufficiently strong cover secured against displacement (Fig. 2). Since covers have to be removed from time to time and they are not always replaced, the responsible supervisor should keep a constant watch to see that openings are not left unprotected. Special attention should be paid to doors leading into the open, and doors leading to unfinished lift shafts. People should be effectively prevented from walking through such openings. A simple barrier that can easily be removed, or under which a man can crawl, is not a reliable safeguard.

Approaches to elevated workplaces

It should be possible to reach elevated workplaces safely and easily from ladders, ladderways or gangways. Ladderways and stairs should not be carried more than two storeys in a single flight. Railings should be provided at the sides of ladderways and stairs and gangways. Landings on working or scaffold floors should be fenced.

1.1.10 Constructional requirements concerning safeguards against falls of persons

(a) Railings and toeboards

Railings should be designed to withstand a lateral pressure of 50 kgf and the posts should be spaced accordingly. Scaffold uprights, boards, metal tubes or metal sections may be used for rails. Only in special cases, where such rails cannot be used, as with brackets scaffolds for tall chimneys, may taut ropes or chains be stretched between the posts. The top edge of the railings should be 1 m above the floor. They should not give way when a person leans against them. If wooden railings are nailed together, the number and length of the nails should ensure a sound assemblage. A poorly or carelessly fastened railing is more dangerous than no railing at all. Where for operational reasons part of a railing has to be removed, it should be so designed that the part can be easily and reliably removed and replaced.

No less important are the design and fixation of railing posts: they also must be able to withstand the stresses assumed for the railing. Good workmanship in the making of the fastenings of railing posts is very important, especially when they have to be placed at or on concrete floors. It is not a good arrangement to attach posts by nailing or recessing to loose or inadequate boards.

With concrete floors, recesses, which in any case have to be provided for the the construction of the shuttering and the laying of the concrete, are preferable. In these recesses both wooden and metal posts can easily be secured by wedges, which can be withdrawn just as easily. Pouring concrete into the recesses afterwards will cause no difficulties. Other methods of fastening are anchoring metal posts in the reinfording steel of the floor, and attaching them to devices let into the floor (screws and bayonet fastenings). All these methods of fastening also allow the railing to be built as a unit that can easily be fitted into place, and enables a floor to be fenced before the walls are begun.

A person may slip between the railing and the floor and experience has shown that a toeboard is necessary to complete the

 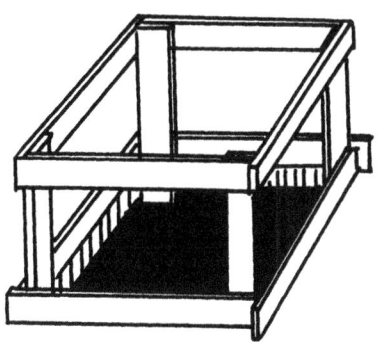

Fig. 2
Protection of floor openings

protection against falls. Its height will depend on whether or not it is also intended to prevent the fall of materials and tools. In this respect official regulations vary, the required height ranging from 15 to 30 cm. Toeboards should have some special arrangement to prevent them from overturning.

The additional requirement of an intermediate rail, halfway between the top rail and the floor, is not universal. The intermediate rail is, however, indispensable when work is done sitting on the scaffold floor, as in painting: the intermediate rail then makes a secure back rest.

(b) Retaining surfaces (horizontal, vertical and inclined hoardings and platforms)

Retaining surfaces, whatever their type, should be so designed that persons, materials or tools cannot fall over their edge or through them. They should not be more than 2 m below the workplace and should be at least 1 m wide. In addition horizontal platforms require a railing at least 1 m high on the edge so that a person falling on the platform cannot roll off it.

Horizontal platforms, in some countries called catch platforms, are most often constituted by outrigger or bracket scaffolds, but even pole scaffolds may be erected for the sole purpose of affording protection against falls. To fit vertical or inclined hoardings is not always easy, and may require some study. Conditions are simple when the building has window openings. Catch platforms can also, like bracket scaffolds, be suspended from hooks fastened to the reinforcement of concrete floors, and so can easily be put up and taken down by tower slewing cranes.

Catch platforms may consist of boards, steel tubes with a strong wire netting, or nets. Hoardings fitted on the edge of a sloping roof should extend at least 1 m above the roof. They should be able to hold securely a load of 150 kgf rolling from a height of 5 m on a roof sloping at an angle of 60°.

(c) Catch nets

Catch nets should conform to official standards as regards quality of the material and mesh size. A net and its suspension device, and also the building part to which it is to be attached, should be able to hold a load of 150 kgf falling from a height of 6 m, without any damage to the net. The suspension device should ensure that the net is securely fastened, and that persons falling into it cannot roll out or be injured by the suspension. Before every occasion of use a net should be examined as to its safe condition. Damaged nets should be taken out of use.

(d) Safety belts and life lines

Safety belts serving to protect persons against falls consist of a body belt, one or two eyes and a life line or belt.

Only officially recognised safety belts should be used. Consequently every belt should bear the manufacturer's name or mark, the year of manufacture and the recognition mark in some durable form. Safety belts should conform to the relevant standards or regulations. Life lines should be kept as taut as possible, and should not be shortened by knotting. They should be securely fastened to a solid anchorage as nearly as possible vertical above the workplace. They should not be led over sharp edges. To avoid slack, measures should be taken or devices used that will keep them at the right length.

For the storage, testing and repair of safety belts and life lines see section 5.

1.2 Excavations

General

Excavations in the building industry are usually either pits for foundations or trenches for piping. In both kinds of excavation accidents due to collapse of the sides are frequent; they are usually serious and sometimes fatal. Although the accidents are similar, their causes differ, as do the measures that should be taken to prevent them. Means of preventing collapse of sides include sloping them and supporting them. The measures to be taken in a particular case will depend on the firmness of the ground. Consequently a person in charge of excavating operations should not only possess the knowledge and experience required for making slopes and digging pits and trenches, but should also be able to judge the firmness of the adjacent ground correctly.

The firmness of the ground

Ground may be divided into three classes as regards its firmness:

- non-cohesive ground, or light soil;

- cohesive ground, or heavy soil;

- rock.

Examples of non-cohesive ground are pure sand and gravel, especially when dry, because they consist of particles that do not stick together and consequently shift. Their firmness and hence their natural angle of repose is very low. This angle depends on the mutual frictional resistance of the grains, which with very fine dry sand which flows easily is practically nil. With uncohesive ground or light soils such as sand and gravel, the slope of the ground, that is the angle to the horizontal at which the side of an excavation is stable, cannot be reckoned to be more than 45°. This angle is formed when the base of the slope equals the depth of the excavation, and the slope is said to be 1 in 1.

In cohesive ground the frictional resistance of the particles is greater. There is also the mutual attraction of the particles, or their capacity to cohere. Cohesion is affected through pores which absorb moisture. The most typical example of a cohesive soil is clay. All soils containing clay can be said to be cohesive. Loam is a mixture of clay and sand, and there is a difference between heavy loam (much clay, little sand) and light loam (little clay, much sand). Loam can be recognised by the yellowish or reddish colour caused by its sand content. Loam may also contain carbon and decomposition products as well as sand and clay. Light loam with much chalk and quicksand is called loess. Clay with much chalk is called marl.

Since in cohesive soils the effects of the frictional resistance of the particles and their cohesive capacity are combined, the natural angle of repose of these soils is greater than that of non-cohesive soils, and is 60°. Thus for a trench 1 m deep, the base of the slop is 60 cm.

Rock has a mineral bonding. There is a difference in solidity between tightly bonded rock that is hard to break up, and usually has to be blasted, and loosely bonded rock that is heavily fissured, friable, slatey or weathered and so is easy to break up. The natural angle of repose of tightly bonded, or heavy rock, is 90°, and for loosely bonded, or light, rock, 80°. Thus with heavy rock the side of an excavation may be vertical.

The equilibrium of forces in the ground may be disturbed by special influences such as loads, vibrations, excavating equipment or traffic. Even if the side of an excavation at first remains intact these outside influences may cause it to collapse from top to bottom at any moment.

Moisture in ground also reduces its firmness, affecting cohesion more than frictional resistance. Consequently cohesive soils are more liable to be weakened by moisture, especially during frosts and thaws, than non-cohesive soils. Further, ground may be weakened by changes in the stratification and structure of the rocks at the site, the superimposition of different types of rock, intercalated stone bands that make breaking up more difficult, and scarcely visible sand veins. Special caution is needed when the ground has been altered by earlier building operations. Old trenches passing near the building site can be dangerous if they are not discovered in good time. On existing buildings, damaged gutters or absence of gutters may have led to changes in the structure of the ground that make safety measures necessary, although in normal conditions the ground could be said to be firm. The firmness of ground should not be judged exclusively with reference to excavations; account should also be taken of subsequent operations and influences; some of these such as the stacking of materials, and the installation of machines near the edge of an excavation, and traffic passing by it, may cause stresses and vibrations that result in collapse if the necessary precautions are not taken.

Great experience is thus needed to judge the firmness of ground correctly in any particular case, and just as much circumspection in deciding on the most suitable safety measures. These measures may include sloping the ground or reinforcing the sides of an excavation. Since conditions on the spot often do not allow the ground to be sloped as much as the depth of the excavation requires, or give rise to disposal difficulties, in most cases the sides of excavations have to be supported.

Digging of pits for foundations and trenches for piping

These operations have to be well and carefully prepared and carried out, and they should be supervised by competent persons.

1.2.1 Protection of existing buildings and measures in the vicinity of existing pipelines

Before and during excavation operations in the neighbourhood of buildings, streets and railways, the necessary measures should be taken to protect the workers, for instance propping up dangerous masses of earth or structures and underpinning foundations.

Dangerous streams should be dammed up or diverted. It should be ascertained whether there are any pipes in the working area that could endanger the workers. If such pipes or their coverings are found the supervisor should be informed so that he can arrange with the owner or the occupier to have them effectively protected.

1.2.2 Excavation and working space

(a) **Excavation**

As excavation progresses, the sides of building pits and pipe trenches over 1.25 m deep should be adequately sloped or properly supported with due regard to the nature of the ground, ground water conditions and imposed loads. This also applies to the heading of pipe trenches (Fig. 3A). If, in spite of sloping, the falling in of masses of earth is to be feared, the sides of pits should be stepped. The steps should be at least 1.5 m broad and not more than 3 m high. In firm settled ground the edges of pits and trenches that are not deeper than 1.75 m may be adequately sloped to a depth of 50 cm. In such cases the edges of pipe trenches can be secured by boards. The boards should be set about 5 cm above the top of the side so that material cannot roll into the trench. At the top of the slope strips at least 60 cm wide should be kept clear.

When the sides of an excavation are sloped as a safety measure the slope should be as follows (Fig. 3B):

- with top soil and non-cohesive light soil that can be broken up with shovels and spades; maximum 45°;

- with moderately heavy soil that can be broken up with a spade, and heavy soil that can be broken up with a pickaxe or a pneumatic drill: maximum 60°;

- with heavy rock that can only be broken up by blasting: maximum 90°.

When vibrations are to be expected, or in the vicinity of the pipe trench the ground has already been excavated in earlier building operations, or other circumstances make it necessary, trenches deeper than 1.25 m should have sloping or properly supported sides (Fig. 3C). Pipe trenches more than 1.25 m deep that

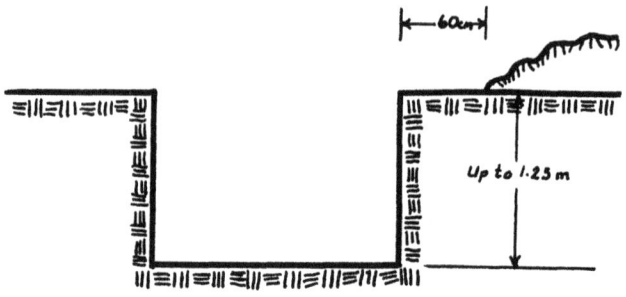

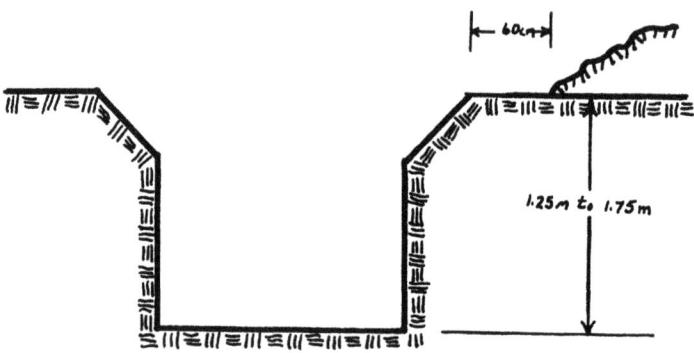

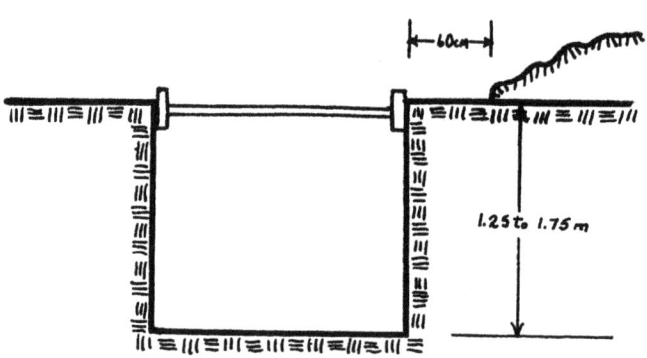

Fig. 3 (A-B-C)
Trench excavation

are excavated by machines such as power shovels and do not have sloping sides should not be entered until the sides have been made safe (section 1.2.4) (Fig. 4, Fig. 5). In ground whose consistency might deteriorate through dryness, penetration of water, frost or the formation of slip surfaces, the sides should be sloped less steeply and measures taken for the safe removal of any water.

(b) Working space

Working spaces in foundation pits

Working spaces in foundation pits should be so dimensioned that between the side of the excavation and any part of the building there is a clear space at least 75 cm wide. In pits with sloping sides this width is reckoned to be the horizontal distance between the foot of the slope and the outside of the masonry or the shuttering. In pits with supported sides it is reckoned to be the clear distance between the supports and the outside of the masonry or the shuttering.

Working spaces in pipe trenches

If pipe trenches have to be entered adequate working space should be provided on both sides of the piping. With depths up to 1.75 m trenches should be at least 60 cm wide, and with greater depths, at least 80 cm. In trenches with sloping sides the minimum width is the width of the floor, and in trenches with supported sides it is the clear distance between the supports.

In trenches with sloping sides the width of the working space is reckoned to be: the horizontal distance between the foot of the slope and the outermost points of the piping, not counting sleeves or flanges; or the bricked-up piping; or the outside of shuttering of piping that is to be concreted or walled in.

When the sides of a trench are supported, the width of the working space is reckoned to be the clear distance between the supports and the outermost points of the piping not counting the sleeves or the flanges; or the bricked-up piping; or the outside of shuttering of piping that is to be concentrated or walled in.

The working space on both sides of a building to be erected should be at least 50 cm if the walls of the building are to have shuttering, or extensive work has to be done on them from the outside.

1.2.3 Supports

(a) General

The following types of support are used for the sides of foundation pits or pipe trenches:

- supports with horizontal boards;

- supports with horizontal boards between bearers that are rammed in or fixed in holes;

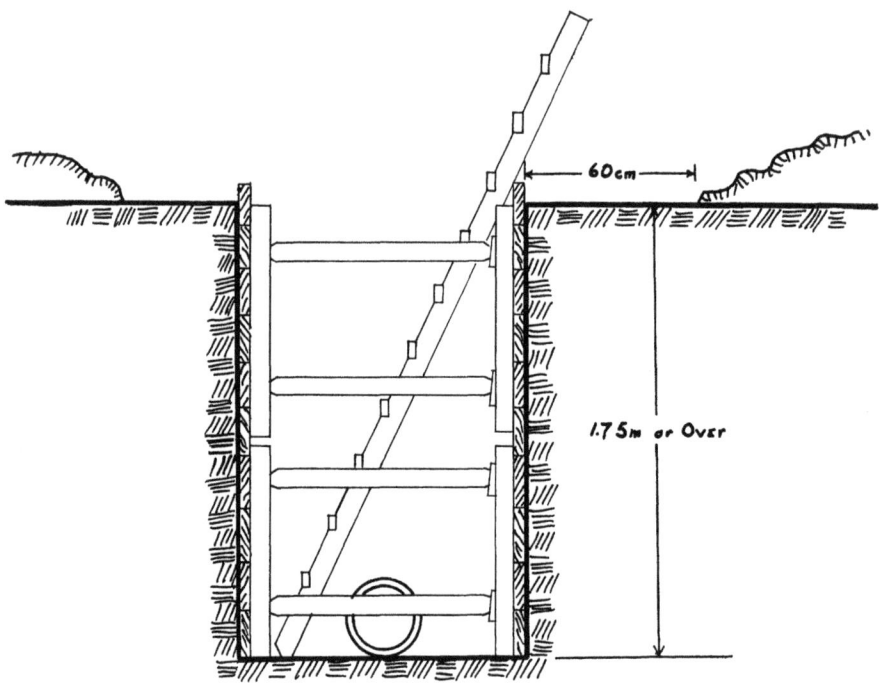

Fig. 4
Trench shoring

Fig. 5
Movable trench shoring

- supports with vertical boards;
- supports with steel bars;
- supports with sheet piling.

In all cases the supports and their parts should be so calculated that they will not be liable to buckle, tilt or bulge under the stresses to which they will be subjected. The calculations should be based on the highest expected load imposed in the most unfavourable circumstances. Account should be taken of bulging, subsidence and vibrations in the ground in and around the excavation, variations in the amount of water present and any disturbances in the ground. Supports should only be made, altered and removed by experienced workers. Boards and wales should lie flat against the earth, and cavities behind supports should be immediately filled in and packed tight with twigs, wood wool or other suitable material. No loose material should percolate through cracks and joins in the supports. Horizontal parts of the supports should not have extra loads placed upon them if this would impair their solidity; this means that working platforms should only be erected on struts if they are secured against slipping. Supports should be constantly watched, and when necessary repaired and extended. Particular attention should be paid to the consequences of vibrations from traffic and working equipment. Inspection is also necessary after length interruptions in the work, prolonged rain or blasting. Supports should only be altered or removed on the orders of a supervisor. Supports should be removed by methods that do not endanger the workers concerned, and not until they have been rendered superfluous by filling in the excavation (section 1.2.4(f)).

(b) **Supports with horizontal boards**

Supports with horizontal boards can only be used when the ground is so firm that it can be excavated to a depth equal to the width of a board without a board having to be inserted. Such supports should be steadily built up as the excavation proceeds; they should keep within one board width of the excavation, or two if the ground is firm. The first boards of the supports should be placed before the depth of the excavation exceeds 1.25 m, or a lesser depth if the ground is loose. The top boards of supports with horizontal boards should extend at least 5 cm above the top of the excavation.

In foundation pits in which supports with horizontal wooden boards are used the following requirements should be satisfied:

- boards should be at least 5 cm thick, with parallel rectangular edges;
- wales should be at least 8 cm thick and 15 cm wide. Boards may not be used instead of wales;
- supports without wales are not permissible. Wales should be supported by at least two struts. Wooden struts should be at least 10 cm thick. With steel struts consisting of an inner tube sliding in an outer tube and two caps, the maximum permissible load with the greatest extention should be calculated with a safety factor of two.

With horizontal supports the joints between the panels should be doubly braced, that is braced on both sides of the join. With boards between 2.5 and 4.5 m long there should be further bracing in the middle of the panel. In fine sand additional measures should be taken to protect horizontal supports against collapse, for instance by affixing long wales. There should be no gaps in horizontal supports, either between the boards or at their joins. Only boards of the same length should be used in one panel.

(c) Supports with horizontal boards and bearers rammed in or resting in holes

In foundation pits with horizontal supports mounted on bearers rammed in or resting in holes the boards should be so long that they rest on at least a quarter of the flange width and are forced against the side of the pit by wedges or other devices.

Struts that lie between the bearers should be secured against slipping out by hardwood wedges, and by special means against falling out: merely hanging them on packing wire is not sufficient. If the bearers cannot be set or driven in to a sufficient depth to give a firm seating, packing pieces should be used.

(d) Supports with vertical boards

These supports may be used when loose ground prevents the use of horizontal boards. They should always keep up with the excavation. The wooden boards should be at least 5 cm thick with parallel rectangular edges. Horizontal wooden beams with a cross section of at least 12 x 16 cm should rest in iron hangers which should have a cross section of at least 10 x 30 mm or a diameter of 16 mm.

If the boards cannot be driven in far enough to give a firm seating intermediate struts must be used.

(e) Supports with steel planks

Lining with steel planks may be used when it does not have to be packed or locked together. The planks should have the same shape over their entire length, and adjacent planks should well overlap when driven in. Bulging or bent planks should not be used. In every phase of the excavation steel planks should be sunk at least 30 cm into the ground. If they can only be driven down to the floor special measures should be taken. The struts should be so placed in straps that the planks cannot be displaced or bend in an undesirable way.

(f) Supports with sheet piling

The effective seating depth of sheet piling should be ascertained and provided with due regard to the ground and water conditions. It should be assembled by tongs, straps or bars. All parts should be strong enough to withstand the maximum expected stresses and to avoid excessive sagging. Boards that do not lie flat should be backed with packing. Struts should only be placed against the tongs or straps.

(g) **Supports with rigid frames or cages**

If pipe trenches are excavated to their full depth with bucket excavators without supports the requirement that unsecured trenches should not be entered can only be complied with when, subject to the observance of special precautions, provisional supports are lowered into the trench to serve as protection while the definitive supports are being installed. This is only possible when rigid frames or cages are used, but for their use to be possible a trench should be excavated with smooth walls which reamin intact until the definitive supports are installed.

1.2.4 Other precautions

(a) **Crossings**

Pipe trenches more than 80 cm wide should be provided with a sufficient number of crossings. They should be at least 80 cm wide, offer adequate support and be so secured that they cannot slip.

(b) **Safety lanes**

On each side of a trench or a pit a safety lane at least 60 cm wide should be kept clear. If conditions on the spot prevent this, precautions should be taken to prevent excavated material from falling back into the excavation.

Material that cannot be stored should be removed.

(c) **Ladders**

Pipe trenches and foundation pits over 1.25 m deep should only be entered and left on ladders, so a sufficient number of ladders should be provided. Climbing in and out over the supports should not be allowed.

(d) **Head protection, life lines**

During work in trenches and pits suitable head protection should be worn (hard hats, see section 4.2). Where in work on earth walls there is a risk of falling, the workers should be roped, for instance at the edges of steep slopes, at elevated workplaces that do not offer sufficient foothold, and when natural conditions (rain, snow, ice) make walking or standing unsafe.

(e) **Protection against falls of rock and collapse of masses of earth - barring dangerous places**

In deep foundation pits the sides and slopes should be carefully examined for loose masses and cleared before each shift, in thaws, after heavy rain, and after any blasting. Work should not be resumed until the necessary precautions have been taken under the supervision of the person responsible for the site.

(f) **Filling in trenches and pits**

Supports should not be removed until they have been rendered superfluous by filling the excavation. If they cannot be removed safely they should be left in place. When trenches with horizontal supports are filled in the boards should be removed one by one, and temporary struts should be used. The struts at the end of panels should be removed last, and then no one should be in front of the panel.

1.3 Underpinning and shoring operations

Underpinning and shoring operations are reckoned to be some of the most difficult and most dangerous building operations. Any disregard of recognised technical or constructional principles can threaten the solidity of the structure to be underpinned or shored, or parts of it, or adjacent structures. If serious injuries to persons and damage to property are to be avoided, the state of the structures should be carefully examined before the operations begin and thorough preparations should be made in which all necessary precautions find a place. Underpinning and shoring operations should only be undertaken by firms possessing the necessary knowledge and experience and equipment, and able to guarantee sound work.

Underpinning operations

1.3.1 Preparations

Underpinning operations should be carried on under the direction of a competent experienced person who should be in constant attendance at the site. The results of the preliminary examination of the structure to be underpinned and any adjacent structure will determine the working method to be adopted. The safety measures required for this method should be set out in a precise plan of operations and timetable, together with lists of the necessary materials and equipment. Before the work begins all the equipment and materials required for safety measures should have been provided and be ready for use. Only in this way can dangerous improvisations be avoided in the operations.

1.3.2 Excavations

No building, not even one in a provisional stage of construction, should be laid bare to the bottom of the foundations or deeper without adequate precautions being taken. If supports are not to be used along the outside walls of the neighbouring buildings a ledge should be left with the dimensions given below as protection against subsidence in the zone of support.

The top of the ledge should be at least 50 cm above the bottom of the foundations but not below the level of the basement floor of the adjacent building. The ledge should be at least 2 m wide and along the edge there should be a slope of not more than 1 in 2. These specifications should be most strictly observed when the ground is excavated with an excavating machine.

The ledge should be removed, and the foundations built, in stages. When the difference in level between the top of the ledge and the bottom of the foundations exceeds 1.25 m the different stages in the excavation should be protected by supports.

1.3.3 Foundations adjacent to existing buildings

New foundations immediately adjacent to neighbouring buildings should be carried down at least to the depth of the foundations of these buildings. If the floor of the new foundations is below the neighbouring foundations these should be underpinned, or if possible be secured by injecting the ground underneath, at least alongside the new foundations. These foundations should be at least 50 cm high and be built in lengths of 1 to 1.25 m. Only sections with a centre-line spacing of at least four times the length should be built simultaneously. In foundations with positive longitudinal reinforcement (ring anchors) the ends of the anchors of the various sections should first be bent upwards. If the reinforcement has to be static, first an unreinforced foundation about 50 cm high and with the bottom level with the existing foundations should be built in sections. When they have sufficiently set the new reinforced concrete structure should be built over the whole length.

1.3.4 Underpinning existing buildings

If the foundation floor of a new building is lower than that of the neighbouring building, the existing building must be underpinned at least alongside the new foundations, or if possible, should be secured by injecting the ground under the foundations.

The underpinning should be built in sections from 1 to 1.25 m long, with axis spacing not less than four times the length. If necessary not only the adjacent wall but also the connected walls should be underpinned. The length of the underpinning of any of these walls will depend on the depth of the underpinning and the type of building.

Effective transmission of forces in the underpinning should ensure that subsidence will not be serious. Each section of underpinning should be completed over the full width in one shift. Door and window openings affected by underpinning should be braced.

The new foundations should be built in sections to keep them abreast of the foundations of the underpinning, so that the floors of the two foundations are at the same level.

While underpinning is in progress the neighbouring buildings should be watched for cracks. Before the work begins any existing cracks should be marked with gypsum for inspection purposes.

To prevent cohesive ground from being weakened by ground or surface water the site should be covered with tarpaulins. Frost damage can be avoided by using heat-insulating coverings.

Shoring operations

If the complete or partial demolition of buildings leaves neighbouring outside walls unsupported, measures should be taken for the security of the neighbouring buildings. If the walls to be supported are not too far apart horizontal struts may be used. For greater distances between the walls, and where horizontal struts are not practicable, vertical shoring or trusses may be used.

In underpinning operations, as for example when basements are added subsequently or alterations are made or parts are replaced, inside and outside walls should be shored and braced. In such operations the greatest caution should be exercised. Special care is needed in examining those parts of the building which will have to support greater vertical loads in the future.

1.3.5 Bracing party walls

(a) Horizontal bracing

With horizontal bracing, wooden crosses with arms from 2 to 2.5 m long rest on round steel sections driven into joins in the walls. Between the crosses bracing beams are inserted and secured with double wedges. The crosses are also secured to the beams with four stays. The stays are recessed into the crosses and fastened to the beams with blocks of wood and screw bolts so as to avoid weakening the cross-section. This arrangement of the stays distributes the pressure and at the same time holds the beams in place while reducing their buckling length. If the bracing consists of a number of beams, they should be linked together with triangular bracing.

(b) Vertical bracing

Where distances between buildings are considerable, or horizontal bracing cannot be used, vertical bracing should be used. Allowance should be made for the horizontal thrust on the inclined bracing. The inclined supports should rest against an upper floor or a braced part of the wall, and the feet should be wedged to prevent slipping.

To secure outside walls suitable trusses can usefully be employed. With this arrangement at about the distance of a storey height above the vertical posts, beams are laid on which the truss with its rather weaker bracing rests to form a sort of strut frame. The tensile stresses are taken by round steel sections, which can be adjusted by means of stretching screws. The trusses are anchored in the masonry of the outside walls with round steel sections.

1.3.6 Shoring and rebuilding walls

When walls are rebuilt it should be remembered that a supporting or bracing structure is to be replaced by one of at least equal strength and stability. This also applies to the removal of walls and floors which form part of the general supporting structures of a building, and serve as horizontal or vertical bracing both internally and for adjacent buildings.

Only sound wood, as dry as possible, straightgrained and in one piece should be used for shoring. The cross-sections of vertical and inclined shoring, as determined on the assumption that there will be a sufficient number of shores and that round timber will be used, should be at least 20 cm in diameter. To rely on experience for constructing and dimensioning shoring requires great practical experience that can only be acquired after many years of difficult rebuilding operations.

Preferably vertical and horizontal bracing should be secured to window and door jambs. This also applies to the rebuilding of load-bearing and partition walls. The arrangement of the shoring should be such that subsequent building operations will not be hampered, and that the shoring itself will not have to be altered. The form and the security of the supports for the vertical bolts and the inclined shores are of particular importance: in any case, these supports should be rigid. Since shoring will be imperiled by concealed cavities and weak ground, the beams should be placed on boards or beams, clamped and braced. Shores should not be set on intermediate floors, arches, etc. unless these are reinforced under the load-bearing points. Care should also be taken that the floor on which shoring rests cannot give way.

For shoring outside walls, of shop fronts, for example, two-row shores with bracing should be used. These support the upright carrying the crossbar, which may be a timber beam or a steel girder, held by hardwood wedges. Then the inner shores can then be set on intermediate floors, arches and the like, if in turn these are adequately supported, or they can be led through floors after holes have been made in them. The sole plates should be let into the floor. Basement walls should be secured by bracing against lateral pressure. Walls below tiers of beams should be shored with inclined supports. The longitudinal bracing of the vertical shores is effected with crossed beams joined together with screw bolts. Rebuilding inside is chiefly carried out by removing supporting and partition walls or by making and closing openings in the masonry. As with outside walls, shoring of inside walls requires the use of sufficient shores joined by bracing and resting on sufficiently large sole plates. If upper floors have to be shored they should be supported in a similar manner, if necessary down to the basement floor.

1.4 Work with concrete and reinforced concrete

General

In work with concrete and reinforced concrete a distinction can be made between precautions concerning the safe design, construction and use of the necessary equipment, and those concerning the production of good-quality sound concrete whether reinforced or not. Persons working with concrete or reinforced concrete are endangered not only by the use of defective equipment but also by the building itself: its stability may be so impaired by poor work that parts collapse even during construction, perhaps while shuttering is being removed. The technical requirements applying to equipment commonly used in work with concrete and reinforced concrete have been dealt with in the section on equipment, section 2, as follows:

Shuttering (forms) and its supports, sections 2.5.12 and 2.5.14.

Concrete mixers, section 2.1.4.

Steel bending and cutting machines, section 2.1.2.

Concrete compactors, section 2.9.10.

Lifting appliances, section 2.3.

Pneumatic concrete conveyors, section 2.4.4.

Defects that seriously impair the solidity of a concrete or reinforced concrete structure or part of a structure may be caused by the use of unsuitable building materials, such as bonding, steel additives, water, concrete additives, or by the incorrect supply, composition or processing of the concrete or reinforced concrete so that the proper quality is not obtained. If these defects mean that the concrete fails to set properly, and its estimated bearing capacity is not reached by the time the shuttering is removed, and if the defects are serious enough, parts of the building may collapse. Workers who are on or under these parts will be seriously injured.

Requirements concerning work with concrete and reinforced concrete

The erection of concrete and reinforced concrete structures requires a thorough experience of this type of work and that appliances and tools are available that will enable the work to be done properly with concrete of uniformly good quality.

1.4.1 Requirements concerning management and workers

Firms doing jobs with concrete and reinforced concrete should have a reliable managerial and supervisory staff familiar with the work involved and equipped with sufficient knowledge and experience concerning the preparation, processing and assessment of concrete. The manager of the firm or a competent representative should be present at the building site while the work is carried on. He will be responsible for seeing that it is done properly and in particular for the following matters:

- the dimensioning of the structural parts according to the plans;

- the proper erection and bracing of the shuttering (forms) and its supports;

- avoidance of overloads, as in the transport of concrete or the storage of building materials;

- the good quality of the building materials used, especially the concrete;

- the proper production, transport, processing and subsequent treatment of the concrete;
- the proper insertion of the steel elements;
- the correct timing of the removal and dismantling of the shuttering;
- on buildings made of prefabricated parts, the removal of shuttering from damaged parts, and the proper erection of the necessary supports for erection operations;
- avoidance of overloads on finished structural parts.

At the building site particulars of all matters affecting the soundness of the building and its parts should be recorded and kept up to date. These particulars should include:

- the times of the different operations such as delivering the concrete, and removing the shuttering and its supports;
- the air temperature and the weather conditions while the different parts of the building were being erected, with special mention of days of frost;
- the names of suppliers and the numbers of the invoices for cement, prefabricated parts, transported concrete, separate additives of different particle size, or ready-mixed additives, with an indication of the part of the building for which they were intended;
- the composition of the concrete in every mixer charge, the weight of fresh concrete, the amount of cement per m^3 of compacted concrete, the type and quality of cement, sieve fineness and particle sizes of the additives, amount of concrete additives, consistency of the concrete and water content of the cement;
- the making of all concrete test pieces, the results of the tests on them and the dates and serial numbers of the tests;
- the results of examinations of fresh concrete, and tests of bonding materials and additives;
- the nature, quality and manufacturer of the reinforcing steel and the results of tests on the steel at the site;
- certification of the proper dimensions and insertion of the reinforcement.

1.4.2 Requirements applying to the building site

At the building site sufficient facilities, appliances and tools should be provided and maintained for the production, use and assessment of concrete to ensure that the concrete is of good quality, in particular facilities, appliances and tools for the following purposes:

- storage of building materials, for instance dry storeplaces for bonding materials, clean storeplaces for reinforcing steel and additives, the latter separated as far as necessary by particle size;
- measurement of bonding materials, additives and water;
- mixing concrete;
- transporting and processing concrete, for example mechanical compactors;
- subsequent treatment, for example coverings or appliances for wetting the concrete;
- testing building materials and concrete, for example trial sifting of additives, consistency tests and verification of the cement content of fresh concrete;
- making and proper storage of test blocks for pressure tests.

According to the nature and quality of the concrete used further requirements could be laid down for appliances, etc.

Building materials

1.4.3 Bonding materials

Only bonding materials that satisfy the quality requirements of the relevant building standards and are so marked should be used. They should be protected against dampness. Vehicles and silos for bonding materials should contain no residues of other kinds or qualities of cement or of other materials.

1.4.4 Concrete additives

Additives should conform to the relevant building standards as regards type and quality. As far as practicable the mixture should be coarse and free from cavities. The coarsest kind should be chosen to suit the mixing, transport, delivery and processing of the concrete.

1.4.5 Other ingredients

Ingredients that serve to improve some properties of the concrete but may adversely affect others need to be tested in concrete for suitability. Chlorides and chloride-containing substances should not be used in concrete.

1.4.6 Mixing water

Naturally occurring water is suitable for mixing provided that it is not rich in ingredients that hamper setting, as is industrial waste water and marsh water.

1.4.7 Concrete and reinforcing steel

Concrete is divided into quality classes according to its test-block stability after 28 days. For these classes there are building standards specifying the uses of each class.

To verify the quality of building materials and concrete on the site, tests are prescribed as follows:

- for concrete, suitability, quality, setting and composition tests;
- for reinforcing steel, verification of the quality mark and the manufacturer's name, and testing of weldability.

Preparation and processing of concrete

The composition of a mixer charge and the cement content in kg/m³ of compacted concrete should be posted up legibly at the mixing point. The weights of the ingredients of the concrete - cement, additives and water - should be indicated.

The ingredients must be mixed in an efficient mixing machine until a uniform mixture is obtained. Only poor quality concrete should be mixed by hand. Mixing machines should only be operated by experienced workers who are capable of seeing that the requirements applying to concrete are respected. During transport concrete must be prevented from separating. Concrete should be processed without delay and should be compacted in the shuttering as completely as possible by shaking, ramming, stamping or banging. Care should be taken that the steel reinforcement is tightly enclosed in concrete. Until it has sufficiently set the concrete should be protected against harmful influences, such as extreme cooling or heating, drying out, cold, running water, chemical attack, oscillations and vibrations.

In cold weather and frost, owing to setting difficulties and the possibility of permanent damage, concrete should have a certain minimum temperature when delivered and should be protected for some time against loss of heat by freezing or drying.

Placing the reinforcement

Before it is placed in position the steel should be freed from dirt, grease and loose rust. When internal vibrators are used for compacting the concrete the upper reinforcement should be so arranged that they can be taken to all necessary places. The steel rods should be fastened to the shuttering by distance pieces so that they are not displaced when the concrete is poured and compacted. The reinforcement should be protected against corrosion by a tight and sufficiently thick layer of concrete.

Shuttering (forms) and its supports

The shuttering and its supports in the way of bearers, timber beams, anchors, etc. should be strong enough to withstand safely all vertical and horizontal stresses. For the construction of shuttering and its supports, the amounting and dismantling of shuttering, and the duration of shuttering, see the sections on shuttering and its supports, sections 2.5.12 and 2.5.14.

1.5 Erection of prefabricated parts

General

The erection and assemblage of prefabricated building parts was formerly confined to wooden and steel structures, but more recently such parts have been made of reinforced concrete. The accident risks specially associated with the operations are as follows:

- risks of falls of persons;
- risks of insufficient solidity of the parts during erection;
- risks in transport, lifting and assembling of heavy structures.

The number of accidents in erecting prefabricated parts is large, and most of them are serious. Although it is important to take suitable measures to prevent these accidents experience has shown that it is difficult to lay down detailed regulations: they are usually confined to general provisions with indications of their purpose.

Planning

It is especially necessary for the erection of prefabricated parts for technical measures to be supplemented by good organisation. The success of safety measures depends largely on whether they have been allowed for at the planning stage.

Another important prerequisite is co-ordination of the work of the builder's office with that on the site, between the planners and the erection supervisors. The dimensions, weight and form of the various parts depend to a large extent on the transport and lifting appliances available for erection, while conversely the statics of erection should depend on the solidity of the parts during the various phases of erection. To avoid risks in transport and erection these operations must be properly provided for in the planning of the building as a whole.

The erection instructions should also be drawn up at the planning stage in co-operation between the planning office and the erection team. Only in this way can it be ascertained in good time what technical measures of assistance and protection should be planned for the subsequent erection. This is specially important in the erection of prefabricated parts in reinforced concrete, where it is often only with great difficulty that attachments can be improvised during the operations. Attachments for use in erection are indispensable for fastening:

- the hooks, etc. of lifting appliances;

- scaffolds, catch nets, gangways, etc.;

- life lines of safety belts;

- other auxiliary equipment required during erection.

If such attachments are provided for in good time they can easily be made with female screws embedded in the concrete of the prefabricated parts to be erected. These can be used for fastening hooks or for anchoring scaffolds. Round steel loops bound to the reinforcement of concrete parts can be of great use in the fastening of hand ropes or tag lines, etc. However, with round steel loops care must be taken that they lie in the line of pull of the crane when structural parts are being lifted. Otherwise neglected additional bending stresses may break the loops.

Site preparations

Preparatory planning should be followed by preparations at the site. When the site is laid out care should be taken as far as practicable to build traffic ways outside the turning and danger zones of lifting appliances. Because of the great weight of most prefabricated parts, the bearing capacity of the travelways of lifting appliances will need the closest consideration. This applies both to the tracks of tower slewing cranes and the paths and stations of motor vehicle cranes. Provision in good time of all necessary scaffolds, appliances and other equipment for the erection operations also forms part of the site preparation as in all building operations. Erection makes very heavy demands on the theoretical and practical knowledge, experience and physical fitness of those engaged on it and these persons should be carefully chosen. Erection operations should also be under competent direction.

Erection

When the necessary conditions have been created by suitable planning and careful preparation of the site it is not difficult to provide adequate protection for erection workplaces: it is then a question of choosing the best safeguards for the particular operations so that the accident risk is kept as low as possible.

1.5.1 Access to the workplaces

For assembling operations sufficient safe means of reaching elevated workplaces should be provided and they should be used by the workers. They may include portable ladders, suspended ladders, fixed ladders and boatswain's chairs. As far as practicable stairways should be extended as the building rises. If workplaces cannot be reached directly from the means of access gangways should be provided; they should satisfy the requirements set out in section 2.7. As an exception, gangways that are used by only one person at a time for reaching a workplace need only be 50 cm wide.

1.5.2 Protection against falls of persons

The most difficult task in the erection of prefabricated parts is always the protection of the workers against falls. A thorough examination of the conditions should be made to ascertain which among all possible methods of protection is the best. Consideration may be given to railings, hoardings, catch nets, safety belts, lifting appliances and boatswain's chairs; for detailed information on these points see section 1.7 on protection against falls of persons.

1.5.3 Solidity of building parts

To ensure that moveable building parts remain intact and secure in the various stages of erection, care should be taken that the necessary rigging, fastenings, struts and bracing have their proper places in the plan of operations, are actually provided and remain in place until the structures in question are self-supporting. If unforeseen changes have to be made in the plan the necessary decisions should be taken only by persons possessing sufficient technical know-how to enable them to judge the stability of the building correctly.

1.5.4 Transport and lifting

The transporting and lifting of heavy prefabricated parts does not create any problem unknown in the movement of ordinary loads. However, in the erection of these parts there are some peculiarities that need attention. Not every lifting appliance is suitable for handling prefabricated parts even if the lifting capacity is adequate. For one thing, prefabricated parts must be landed within millimetres of their intended position, and for another, to avoid jerks, the lifting speed should not be high. The free length of rope should not be too great so that the load cannot swing wide out of control as it might if the distance between the top pulley and hook was excessive. It is useful to guide the load with a hand rope or tag line.

Before prefabricated parts are slung they should be examined to see whether they have suffered any damage in transport to the site. Damaged parts should only be erected if the damage has not impaired their solidity and stability. Parts placed in position on the building should not be released from the lifting appliance until they have been secured against slipping, overturning and falling.

1.6 Work at and on roofs

Work at and on roofs comprises covering and recovering roofs including fixing laths and boards, work on roof structures and chimneys, work on cornices, the installation of gutters and snow barriers, the erection of lightning conductors and aerials, and painting, cleaning and repairing on and over roofs.

The causes of accident are as varied as the types of roof, the nature of the covering and the kinds of work done; the means of prevention also vary. The types of accident include falls from the edges of flat roofs, and falls from sloping roofs when a worker loses his balance, or wears unsuitable footwear, the roof is wet or icy, or a roof ladder slips down or shifts sideways. There are also the accidents on fragile roofs, both flat and sloping, typified by glass and asbestos cement roofs. But even reinforced concrete slabs as used for hall roofs will have inadequate bearing capacity if the joins are weak and the slabs can slip off their supports. There are other accidents apart from falls, for instance electrical accidents due to overhead lines, and accidents with lifting appliances.

General requirements

Persons who suffer from giddiness, epilepsy or other ailments prejudicial to safety at work should not work on roofs. Before work is begun at or on a roof precautions should be taken not only against falls of persons but also against falls of building materials and tools, both outside and inside the building. Where conditions on the spot or special circumstances prevent the taking of collective precautions, the worker on the roof should always be equipped with a safety belt and life line that will enable him to secure himself, but even for work of short duration a man should stand by to render aid if necessary.

When protective devices such as railings, hooks and gangways are permanently installed on a roof they should be inspected by a competent person before every occasion of use. Since such devices are particularly subject to corrosion it cannot always be assumed that they possess the necessary strength to fulfil their purpose.

If no outside scaffolds are erected for building or rebuilding operations from which the roof can be reached safely, safe access should be provided from ladders or attics.

Footwear suited to the roof material should be worn when walking on roofs; such footwear should be provided for the workers, subject to suitable purchasing arrangements. Even for small maintenance jobs the workers should be precisely instructed as to their behaviour and the precautions to be taken. Supervision should ensure that the instructions are obeyed.

Covering or recovering roofs

If work at gutters (such as changing gutters) has to be done during roofing or reroofing new or existing buildings the following precautions should be taken before the work begins.

About 1 m below the gutter a scaffold with a close-planked floor should be erected. Along the outside edge of the scaffold a hoarding should be erected; it should be at a distance of at least 60 cm horizontally from the edge of the roof or the gutter and extend at least 60 cm above it.

To prevent falls of persons, this hoarding, from the height of the gutter to the scaffold floor, should consist of boards laid close together or strong wire netting with a mesh of not more than 5 cm. Above the level of the gutter the hoarding may have gaps up to 25 cm wide so as to reduce the surface exposed to wind pressure. If roof work on existing buildings does not involve any work at the gutter or the cornice it will not be necessary to erect a scaffold, but instead a continuous hoarding at least 60 cm high should be erected along the gutter. This hoarding and its fixings should be strong enough to hold safely any falling persons or building materials. Authorisation of the hoarding may be made to depend on drop tests in which loads of 2 x 75 kg are dropped from a roof sloping at 60°. The supports of the hoarding should not be more than 1.5 m apart, and it should always be secured by two sets of fastenings. Workers erecting and dismantling the hoarding should wear safety belts with life lines.

Where scaffolds are erected for work on the shell of the building or for work on the facade they can usefully serve for roof work provided that they are suitably adapted. To prevent them from being dismantled prematurely agreement should be reached between the various contractors. For work on the edge of the roof, for instance making the cornice or installing or changing gutters, scaffolds as described in prior sections can also be used. If, however, in addition work has to be done on the roof only such roofers' suspended scaffolds may be used as afford protection against falls from the roof, that is, scaffolds to which a hoarding can be fitted.

Inspection, cleaning, small repairs and other short-term work

On sloping roofs short-term work should only be done from roof ladders or scaffolds, which should be effectively secured against slipping down and shifting sideways; if ladders cannot be securely installed, the workers should be protected by safety belts and life lines, or suitable equipment for preventing falls should be used. Life lines should be anchored to a solid part of the building. The anchorage should be so chosen that the line is as short as practicable. For requirements concerning the specifications of safety belts and life lines see section 5.1.

Work at or on fragile roofs

For making and repairing fragile roofs such as glass and corrugated asbestos cement roofs, passageways and workplaces with adequate bearing capacity should be built. Stepping on fragile roofs should be forbidden.

Passageways and workplaces built over a fragile roof should be at least 80 cm wide, and secured against slipping and shifting. They should afford a safe foothold, and it must be possible to step on them directly from a ladder or other means of access. On corrugated asbestos cement roofs the passageway over a finished roof need only be 40 cm wide.

The safety measures required in section 1.6 against falls of persons and materials over the gutter are applicable here. At points where fragile roofs are entered or left there should be notices worded "On this roof walk only on the gangplanks". This notice will also serve as a warning to persons unaware of its fragility who have occasionally to go on it for inspections, etc.

1.7 Work on chimneys

General

This section covers only independent factory chimneys and does not include house chimneys. Work on chimneys includes erection, repair and demolition. Usually each of these operations involves work at a great and dangerous height and the main accident risk is falling. Falls of persons are not due solely to lack of precautions against them; any defect in equipment such as scaffolds and lifting appliances can be responsible for a fall, especially if the equipment is not solid enough or has not been installed in a workmanlike manner. Naturally, too, persons working at the foot of the chimney, operating lifting appliances or transporting materials, are endangered by falling objects, and even the smallest objects are dangerous when falling from a great height because of their kinetic energy.

For these reasons work on chimneys is among the most dangerous of all building work.

Competent direction and suitable employees

The direction of work on chimneys requires special technical knowledge and experience that can only be acquired in a specialised undertaking. Thus general knowledge of building operations is not sufficient qualification for the responsible direction of chimney work. Similarly, special qualifications are needed for chimney masons, and above all they must be suitable for work at great heights. The mere climb to and from the workplace requires experienced workers not subject to dizziness, and consequently every worker should be examined to see whether he possesses the physical and mental qualities that are indispensable for chimney work.

Installations

1.7.1 Work and protective scaffolds

Among the scaffolds used for chimney erection one may distinguish between inside work and protective scaffolds, mobile suspended scaffolds and outside bracket scaffolds.

(a) **Inside scaffolds for chimneys**

These may be work or protective scaffolds for the erection, repair or demolition of independent chimneys. As a rule they are so-called bearer scaffolds whose bearers can be displaced with respect to each other and consist of rolled sections, steel pipe or lattice girders. These bearers should be let into the wall sufficiently to obtain an adequate supporting surface. As the building work progresses the scaffolds should be raised so that the floor is always at least 25 cm below the top of the chimney so that the wall above it affords adequate protection against falling outside. With these inside scaffolds the floor immediately below should be left in place to serve as a protective scaffold.

The strength of every bearer should be certified.

The covering of removable bearers should extend over at least one third of the clear width of the chimney.

(b) **Inside suspended scaffolds**

Since the shifting of inside work and protective scaffolds always involves risks, so-called climbing scaffolds have been developed that can serve as both work scaffolds and protective scaffolds. They are raised and lowered by built-in lifting appliances and so can follow the progress of the work. Shifting these scaffolds is much simpler and safer because the floor and the bearers need not be taken up and relaid.

The use of climbing scaffolds in chimneys with a clear width of 3 m and more is economic, and also advisable for safety reasons. Their strength and stability need to be certified in every case. The equipment required for transporting material should be built into the scaffold so that the supporting beam or trestle does not have to be moved when the scaffold is moved.

(c) **Outside bracket scaffolds**

As a rule brackets should only be used for maintenance work on chimneys, and only exceptionally for raising their height when circumstances make it unavoidable. An important feature of these bracket scaffolds is their double suspension as a safety measure. Every bracket should have two hooks for suspension, such that each alone can bear the load. The brackets are suspended from two wire ropes or two tested chains laid round the chimney. In no case should they be suspended from wall hooks, hammered-in clamps or chimney rings. The ropes or chains carrying the brackets should be secured against slipping by means of hammered-in hooks.

See section 2.5.10(b) on bracket scaffolds.

Chimney scaffolding should be erected from the bottom upwards unless lighting platforms or passageways at the top enable the bearers to be brought up. After erection the bearers should be left in place for the subsequent dismantling.

(d) **Mobile suspended scaffolds and working platforms**

Mobile suspended scaffolds or mobile working platforms are often used for maintenance work on chimneys and the erection of independent inside pipes in acid columns and for linings. The floor of these scaffolds or platforms should be adjustable to the different diameters of the chimney. Platforms should be raised or lowered by simultaneous action of the lifting appliances. The suspension rope of the lifting appliance should satisfy the same general requirements as the suspension devices of other lifting appliances. See section 2.3.1(c).

1.7.2 **Lifting appliances**

(a) **General**

The winches and suspension devices of lifting appliances used in chimney work are subject to the same requirements as those described in section 2.3 for lifting appliances in general. Lifting appliances for internal transport should only be power-driven when the clear width of the chimney at the workplace exceeds 1.5 m. If the width is less there is a risk that the load may strike against the scaffold and so imperil the workers on it. Support beams or gallows should have at least a third of their length inside the chimney and be firmly secured. The tip of wooden gallows should be at least 8 cm thick. If the pull on the gallows is one-sided the top should be secured by a counter rope. Beams, outriggers or gallows required for hoisting building materials should not be fastened to the ropes or chains carrying the scaffold.

(b) **Protection of loading and unloading points**

Over the lifting appliance and the entrance to the chimney, a protective roof of planks with a layer of sand, ashes or fascines should be installed. The rest of the danger zone should be fenced off and indicated by warning notices. The hoisting opening in the inside scaffold should be protected against falls of persons and materials by toeboards and railings at least 80 cm high. The railings may be so made that they can be removed to allow the bucket to be swung in and out if conditions so permit. However, the railings of the top landing should only be removable if this is necessary for operating reasons, for example if the chimney has a small diameter. With larger chimneys swinging out and setting down the load is attended by difficulties because it cannot be swung out without a great effort. It is difficult to fence the hoisting opening if the space is confined, and accordingly it should be protected against falls in another way. For this purpose flaps in the work or the protective scaffold have proved useful. These flaps or trapdoors should be so made that they close automatically after the passage of the bucket. If these automatic flaps or doors are fitted, the prescribed edge protection (fencing with toeboards) can be dispensed with.

(c) Buckets, etc.

Building materials, etc., should be transported in containers or devices from which they cannot fall out. The containers should not be overloaded. Care should be taken that the loads do not catch in the hoisting openings in the scaffolds. For transporting stones in chimneys with a large diameter lifting gear with a tray hung underneath has been found useful. In no case should stone be carried in rope slings.

(d) Carriage of persons

Carriage of persons on the lifting appliance should be forbidden. In special cases subject to the approval of the competent authority they may be carried but only if certain precautions are taken which should be specified in detail in the written authorisation. See also sections 2.3.9 and 2.3.10 on the carriage of persons on lifting appliances at building sites.

Climbing chimneys

There is a difference between climbing columns of rungs and climbing without rungs.

1.7.3 Climbing chimneys with columns of rungs

Generally the construction of rungs, ladders, protective hoops and galleries at the top of the chimney is governed by regulations, which lay down specific requirements concerning their characteristics and the method of placing them. The way they are placed depends on the height and the clear width of the chimney. In many cases both outside and inside columns of rungs are required. On chimneys over 40 m high protective hoops should be provided at 3 m intervals: these enable the climber to rest. Galleries, usually at the top, enable the state of the chimney to be inspected from time to time. Instead of a top gallery a lighting platform may be required in the interests of the safety of air traffic.

If a column of rungs is built on a chimney they should be tested before they are first used. They should not be used if they are only knocked in. Rungs that are embedded subsequently should not be used until it has been ascertained that the mortar has set.

1.7.4 Climbing chimneys without columns of rungs

Chimneys without an outside column of rungs should only be climbed by means of portable ladders. The ladders should be equipped with distance pieces that cannot be displaced, and if they are extension ladders they should be telescopic. The bottom ladders should rest on a firm immovable base.

Demolition of chimneys

Chimneys may be demolished by overturning or blasting. The masonry should only be holed at ground level. Before holing begins the state of the chimney should be carefully examined. The ways of escape should be decided and the workers informed accordingly. During holing the chimney should be watched attentively by a competent person. As soon as a crack appears the holing should be carried on with special care.

It is preferable to demolish chimneys by blasting rather than by overturning when circumstances permit. The blasting should only be done under the supervision of an authorised experienced blaster.

Other precautions

1.7.5 **Eye and respiratory protection**

As far as practicable the chimney should be put out of action before work begins at the top of a chimney. If work cannot be avoided while the chimney is in action protective goggles should be worn if there is danger from ash; if there is danger from carbon monoxide respirators with the appropriate filter or airline respirators should be worn; for other dusts, gases and fumes respirators with the required filter should be worn.

Nuisances and risks can originate not only from a chimney undergoing repairs but also from neighbouring chimneys and industrial plants; in these cases also the precautions described above should be taken.

See also section 2.4 on personal protective equipment.

1.7.6 **Head and foot protection**

See sections 4.1 and 4.5 on personal protective equipment.

1.8   **Demolition**

General

The frequency and severity of the accidents that occur during demolition work place it among the most dangerous of building operations. Accidents are mianly caused by persons falling from high, unprotected workplaces, or through openings or by falling objects, but, above all, by the unexpected collapse or overturning of the building, or parts of it. Splinters, sharp objects, nails, etc., cause many cuts and punctures.

Structures such as buildings and plant can be demolished by different methods - knocking down, taking to pieces, pulling down, pushing down, and blasting. By knocking down is meant taking down masonry or concrete course by course, and by taking to pieces, the dismantling of wooden, steel or reinforced concrete structures.

All demolition operations should be directed by a competent person, who should possess ample experience of the type of demolition to be undertaken and of its difficulties. Accordingly, he should have a satisfactory knowledge of methods of erecting different kinds of buildings - reinforced concrete, steel framed, wooden, prefabricated - so that he can recognise the constructional peculiarities of the building to be demolished and decide on the measures to be taken. In addition there should always be a person in charge at the site to give any instructions that may be required and to see that they are obeyed.

The persons employed should be suitably qualified for their work. Young persons should not be employed on particularly dangerous demolition operations, and should be restricted to low levels or to clearing up. All workers should be obliged to wear hard hats.

Before demolition operations begin, the building to be demolished and adjacent structures should be thoroughly inspected by the competent person directing the operations. The results of the inspection will determine the demolition method to be used, and the director will prescribe the safest method of working in full detail. Among other things this method should ensure the stability of parts to be demolished and adjacent parts in every stage of demolition so long as they remain standing. This requirement may make it necessary to take special measures to secure parts by bracing or shoring when they lose their stability by the demolition of parts supporting them or keeping them in place. Serious accidents can happen if measures are not taken in good time to secure alcoves, balconies, cornices, arches, roof trusses, etc., against inadvertent or premature collapse.

In all stages of demolition passageways should be kept clear. Floors and stairs should not be overloaded with rubble. So long as stairs are in use the bannisters should not be removed unless they are replaced by temporary but properly secured railings. When removing boards from floors built on joists or girders, if persons work or move on the joists or girders they should be covered with sufficiently strong and undisplaceable planks.

If danger zones are created during demolition, they should be barricaded, suitably indicated by warning notices or guarded by a suitable person. This job of guarding needs great concentration and the men chosen for it should not be distracted by other duties.

Before a person throws down demolition material or anything else he should make sure that no one is underneath or likely to be struck. A loud warning should also be given. But the thrower himself will be in danger if he can be caught by the thrown material and pulled down with it. Consequently he must also make sure that the material cannot catch in his clothes.

Knocking down and taking to pieces

Structures should only be knocked down or taken to pieces from a safe place, and if necessary, scaffolds should be erected to provide such places. Walls should not be stepped on when they are being knocked down, or beams or girders are being thrown down or tipped over. Work should only be done from tops of walls if they are more than 50 cm wide and not more than 3 m high. If conditions allow, workers knocking down walls should be roped or protected against falls by some other suitable means.

If panelled walls are to be demolished the masonry should be taken down first. If structures under ground level (foundations, etc.) are to be knocked down any adjacent unstable ground should be secured against falling in. Care should be taken that the panelled wall is still stable and holds together.

Pulling and pushing down

As a rule buildings should only be pulled down when they are in a conditon that makes it inadvisable to knock them down. For pulling a building down sufficiently strong wire ropes should be used, and they should be long enough to enable them to be pulled from a safe distance. It is good practice to fasten the ropes from travelling mechanical ladders, which will provide the workers with a safe stand. Pulling should not begin until everybody is out of the danger zone. This zone should be made so large that even if a rope breaks persons will not be endangered by the end whipping round.

Structures should not be demolished by undermining or cutting through them because they might collapse prematurely. Nor should bars or racks be used for demolition because this might make the wall fall on the wrong side and so cause danger.

Blasting

The blasting of buildings or parts of buildings makes heavy demands on the knowledge and experience of the blaster. Experience of blasting in underground construction or in quarrying is not enough, and the blaster should show experience of blasting buildings. Among other qualifications he should have served for a long time as assistant to a blaster in blasting buildings. For the rest all precautions customary in blasting should be taken. The shots should be covered so as not to endanger the neighbourhood. When loading the shotholes and fixing the size of the charge account should be taken of local and structural conditions, and it may be found desirable to divide the structure to be blasted into a number of sections. When vibrations must be avoided because of the instability of the parts to be blasted the shotholes should be drilled with electric, instead of pneumatic, drills.

Special attention should be paid to barricading the area round the firing point before the shots are fired so that persons do not enter the area without authorisation. As a rule, in blasting buildings the shots should be fired by electricity.

Before the firing point is freed after firing, that is before the barricades are removed, the blaster should personally satisfy himself that no dangerous conditions have been created by the blast. Any danger zone found should be barricaded until the next round.

Demolition of structures with special risks

The nature of the structure to be demolished may make special measures necessary. Before demolition operations are begun in works making, processing or storing explosives, chemical works, or gasworks, or in dangerous proximity to gases, the requisite precautions should be decided in consultation with specialists.

When fusion welding is done on structures coated with lead paint, lead fumes are formed, and consequently the welders should wear respirators, with suitable filters.

Isolated chimneys should only be knocked down or overthrown by persons with experience in chimney construction. Chimneys should not be demolished from the top of the wall. Inside work scaffolds or outside bracket scaffolds should be erected for the purpose. As demolition proceeds these scaffolds should be moved so that their floor is always at least 25 cm but not more than 1.5 m below the top of the wall. If demolition material is to be thrown down inside the chimney an opening should be made in the wall through which the material can be removed. While it is being removed, work above should stop. If conditions allow, chimneys can also be demolished by being holed or blasted and then the direction of fall should be fixed in advance.

1.9   Other operations

Painting and spray-painting

1.9.1 General

Fire and explosion risks may arise from the composition of lacquers, paints, impregnating materials and solvents used in painting and spray-painting; there may also be serious health risks. Most of the products are marketed under trade names which do not reveal the composition, etc., and the risks may be hidden. In some countries the manufacturer is legally obliged to give appropriate information about risks on the containers but experience shows that the substance is often poured into a different container which is not marked. Sometimes also the recommended precautions are not taken in spite of clear and specific information.

1.9.2 Building regulations

(a) **General**

When the materials used are flammable the rooms in which they are processed are deemed to be rooms with a fire risk. If the materials are lacquers and solvents with a flashpoint below 21°C, or if lacquers and solvents with a flashpoint of 21°C or over are used and warmed, the rooms are deemed to be rooms with an explosion risk. Accordingly, in such rooms there should be no source of ignition such as sparks from machinery, stoves, open fires or unprotected lights.

Special requirements for protection against fire and explosion are applied to the situation and construction of these rooms and to the equipment they contain, such as spray stands, spray booths and drying chambers. The rooms should also be equipped with plenum and exhaust ventilation to ensure that the threshold of danger to health (or the maximum permissible concentration of harmful gases and dusts) is not reached at any time.

(b) **Special installations**

The strict requirements for protection against fire and explosion which apply in fixed workplaces (touched on in the preceding paragraphs) may be relaxed when the work cannot be done in rooms set aside for the purpose. Local or operational conditions may mean that lacquering or spraying points cannot be isolated as in painting or spray-painting of buildings. However, if temporary rooms for this purpose are installed inside a building the full requirements are generally applicable.

Exceptions may be allowed when all the following requirements are satisfied: not more than 20 g lacquer should be used per m^3 air space, not more than 5 kg in an eight-hour shift, and the room in which the lacquering is done should have a cubic capacity of at least 30 m^3 and an area of at least 10 m^2.

In such cases the precautions should be appropriate for conditions on the spot. They should include the following: within a distance of 5 m from the workplace there should be no stove, open fire, unprotected light or machine liable to give off sparks. The heating should be designed so that mixtures of solvent vapour and air cannot ignite on it. The electrical installations should comply with the requirements for rooms with a fire risk or with an explosion risk according to the flashpoint of the lacquers and solvents used. There should be a notice prohibiting smoking.

1.9.3 Operating regulations

(a) **General**

In each case the employer should ascertain the flashpoint of the lacquers, paints and solvents used and any harmful ingredients. Only enough lacquers and solvents to meet the needs of half a shift should be kept at workplaces; they should be kept in unbreakable, securely closed receptacles. Receptacles should be removed from workrooms immediately they are empty.

Passage and escape ways should be kept clear. Objects and clothing not required for the work, even working clothes, should not be kept in workrooms. Clothing badly soiled with lacquer residues and solvents should not be worn.

Spraying plant for oil lacquers and other lacquers whose residues tend to ignite spontaneously should not be connected to the same exhaust installation as spraying plant for nitrolacquers. Oil lacquers and other lacquers whose residues tend to ignite spontaneously should not be sprayed alternately in the same plant with nitrolacquers unless it is thoroughly cleaned before each changeover.

Oxygen and other flammable gases should not be used as atomisers for spraying. After the close of work a responsible person should make the rounds to see whether there is any fire or explosion risk in the spraying rooms.

Spray-painting and similar work should be done only by reliable persons over 18 years of age who are familiar with the work. Apprentices over 16 years of age should only be employed on such work for training purposes and under superivsion.

(b) <u>Cleaning</u>

Lacquering places and floors should be kept clean and clear of lacquer residues. Cleaning the floor can be facilitated by first washing it with milk of lime. Paper can be used if it is removed every day. Lamps should be kept clean so that the radiant heat cannot ignite any lacquer residues. Cotton waste, etc., in use should be kept in tight-closed incombustible containers bearing the inscription "flammable". Used cotton waste and the like should be kept in tight-closing, incombustible, marked containers outside the workrooms. The containers should be protected against heat and emptied daily. All residues should be safely destroyed as by careful burning in a pit away from buildings, and with the building in the lee of the wind.

(c) <u>Smoking and handling fire and open lights</u>

Smoking and handling fire and open lights in workrooms should be forbidden. The prohibition should be displayed on durable and conspicuous notices.

(d) <u>Health protection</u>

Adequate ventilation should be provided when workpieces are being removed, or cleaned with solvents, or paint is being removed from workpieces with solvents. If the workpieces are cleaned with solvents by hand, solvent-proof, impermeable and incombustible gloves and aprons should be provided and used. Persons employed on degreasing, cleaning and painting should be provided by the contractor with suitable washing facilities and means of protecting and caring for the skin. For spray-painting, air-line respirators should be worn unless it can be shown that the concentration of solvent vapour in the sprayer's breathing zone is below the threshold of risk to health. For short jobs filter respirators with air-contaminant filters and activated carbon can also be used. Filter respirators with only cotton wool, sponge or colloid filters should not be used for painting because they do not retain the solvent vapour. For spray-painting without exhaust ventilation, for example when portable spraying appliances are used in rooms, airline respirators or respirators with air-contaminant filters and activated carbon should be worn.

Workers exposed to heavy soiling by paints, etc., should be provided with suitable working clothes and gloves. The undertaking should be responsible for cleaning the working clothes.

Food and drink should not be kept or consumed in workrooms.

See also section 3 on dangerous gases and materials, section 4 on personal protective equipment, and section 9 on occupational skin diseases.

1.9.4 Fire extinguishers

In painting rooms a sufficient number of suitable authorised hand fire extinguishers should be kept in readiness at safe, easily accessible and conspicuous places. In addition, means of extinguishing burning clothing such as drenchers should be kept ready for use at suitable places. The drenchers should be protected against paint mist and dust. If the extent of the work so requires further suitable equipment for extinguishing fires should be provided outside the workrooms. Whatever else the contractor does at the outbreak of a fire, the fire brigade should be immediately notified of the outbreak.

Gas welding and cutting

1.9.5 General

Fusion welding and cutting are processes using the energy latent in the gases. For their use, highly developed appliances are necessary, and they should only be operated by competent persons with adequate experience and training. If this requirement is not satisfied, accidents must be expected that will threaten the health and life not only of the welder, but also of his fellow workers. Such accidents may be due to a variety of causes - the risks inherent in the gases used, faulty construction or operation of acetylene generators, improper storage of the carbide required for the acetylene, storage and handling of gas cylinders contrary to the regulations and technical defects in welding equipment. Further, the work is always attended by fire risks. This brief review of the chief accident risks in gas welding and cutting is enough to show how very important it is to comply strictly with every safety requirement.

1.9.6 Fuel gases

The gases used for fusion welding and cutting are acetylene, hydrogen, coal gas and liquefied petroleum gases which are mixed with oxygen in the mixing chamber of the welding apparatus and ignite through the mouthpiece of the burner. They burn at very high temperatures, for example about 3,000°C with an acetylene-oxygen mixture, and about 1,800° with a coal-gas-oxygen mixture.

(a) <u>Acetylene</u>

Acetylene is somewhat lighter than air. In air it burns with a very sooty flame if the oxygen supply is insufficient to ensure complete combustion. If, however, it is mixed in the right proportions with air or oxygen, it burns completely at a high temperature. Within certain limits mixtures of acetylene and air or oxygen are explosive; with air the explosive range is between 2.3 and 80 parts acetylene by volume, and with oxygen, between 2.3 and 93 parts acetylene. This means that such mixtures, which can form unnoticed, are always an explosion risk. They can ignite at the relatively low temperature of 325°C. Ignition may be caused by a dropped iron object, by a hobnailed boot, a spark from an electric hand drill or switch, or a hot stovepipe.

Another important property of acetylene is its tendency under slight over-pressure to decompose into its components, hydrogen and carbon, with the liberation of great heat which can very quickly lead to high pressures and hence to explosions. To prevent such happenings, fusion welding the maximum permissible pressure has been fixed at 1.5 kg/cm^2 at 1 atm. gauge. The higher pressure in acetylene cylinders is governed by special physical conditions and consequently the cylinders are subject to special regulations. If acetylene comes into contact with copper or alloys with a high copper content the extremely explosive copper acetylide may form. Consequently no parts of acetylene plants should be made of pure copper or alloys with more than 65 per cent copper.

(b) <u>Hydrogen</u>

Hydrogen is one of the lightest gases, 14 m^3 of it having the same weight as 1 m^3 of air. It is colourless and odourless, and burns in air with a slightly luminous soot-free flame. Mixtures of hydrogen and oxygen in the range 4.5-95 parts by volume are explosive. The range is much the same for mixtures with air. A mixture of two parts hydrogen to one part oxygen by volume is commonly called oxyhydrogen gas.

(c) <u>Coal gas</u>

Coal gas is considerably lighter than air. Owing to its high carbon monoxide content it is toxic. Mixed with air in the range from about 5 to 36 parts gas by volume it is explosive. Mixtures of coal gas and oxygen are no less dangerous.

(d) <u>Liquefied petroleum gases</u>

Liquefied petroleum gases (propane, butane and their mixtures) in gaseous form are about twice as heavy as air, and hence tend to penetrate into sunken places such as pits and cellars and displace the air: remaining in such places when they are filled with the gases may lead to fatal accidents as a result of oxygen deficiency.

Mixtures of liquefied petroleum gases with air are explosive in the range from 2 to 9.5 parts gas by volume. The risk of the gases igniting is particularly high because they accumulate near the ground.

(e)   Oxygen

Oxygen is somewhat heavier than air. It is not itself flammable but no normal combustion process is possible without it. Oxygen accounts for 21 per cent of the atmosphere and is the basis of all combustion processes in everyday life. In pure oxygen the combustion of burning or glowing substances is very active. Combustion processes are considerably accelerated when the oxygen content of the air is increased by only a few per cent. Consequently oxygen should never be used to improve the ventilation of rooms, tanks or the like. Pure oxygen can cause oils and fats to ignite.

1.9.7 Acetylene generators

(a)   Carbide and its storage

The acetylene used as fuel gas for welding is either delivered in cylinders or manufactured on the spot in generators. The carbide (CaC.) needed for the purpose which can be obtained in various grades of coarseness should be kept in hermetically sealed containers because atmospheric humidity causes gasification at the surface. After they have been emptied carbide drums should be reclosed and the lids should fit tight to prevent the penetration of atmospheric humidity.

In full containers acetylene forms even if there is only a little humidity, and in some circumstances when mixed with air owing to faulty transport arrangements it can be ignited by frictional sparking. Care should be taken when opening the containers; steel hammers and chisels should not be used for this purpose. In workrooms where acetylene generators are operated only the estimated daily requirement of carbide should be kept. Empty containers should be removed from the rooms.

(b)   Types of acetylene generators

A distinction can be made between fixed plant installed in special generator rooms and transportable plants for erection purposes as used especially on building sites. Generators can also be classified, according to the principle on which they work, as water-flow generators, contact generators, falling carbide generators with a fixed bell, and generators with gasification under a moving bell. In all the apparatus carbide and water are brought together and the resulting acetylene is stored. They therefore consist of a chamber in which the carbide is gasified, a chamber in which the acetylene is stored, and accessories such as pressure gauges, regulating valves, safety vales, stop cocks, purifiers, and backfire preventers. The gasification chamber and the storage chamber are often combined especially in small transportable generators.

As a rule national regulations require the manufacture of acetylene and the storage of carbide to be notified to the competent authority, and require generators to be of an authorised type. Generators should be built, operated and maintained in accordance with recognised technical rules. They should be equipped with a purifier to ensure that the gas will be sufficiently dry and clean for use. They should also have arrangements that effectively prevent the penetration of oxygen or air to the generator and render a backfire harmless. Every generator should bear a plate at a conspicuous place giving the manufacturer's name, the year of manufacture, the factory number, the maximum permissible charge of carbide in kg, and the authorisation number of the type test.

(c)   **Installation of acetylene generators**

Particular care should be taken in installing transportable generators. Workrooms in which they are installed should be well ventilated and have at least 60 m³ air space and 20 m² floor area for each generator. Generators should be at least 3 m away from open lights and stoves. Only one day's supply of carbide should be kept in the rooms and only one container should be opened at a time.

Acetylene generators may be used in the open air only when there is no danger of freezing.

(d)   **Transport of itinerant generators**

Within the undertaking acetylene generators should be transported fully assembled to their places of use; gasification should be interrupted and if necessary gas discharged so that the quantity in the chambers is as small as possible.

If a generator cannot be transported with its generating water and hydraulic seal, it should be taken to pieces to avoid frictional sparking of iron parts caused by jolting in transport. Before it is taken to pieces the generator should be emptied of gas and cleared of sludge, and the water should be run off. It should be taken to pieces carefully, and then all parts should have carbide residues and sludge cleaned off, and gas residues should be removed by filling it to overflowing with water.

When generators are being emptied of gas, taken to pieces and cleaned, and as far as possible also while being transported, every care should be taken to see that there is no open light or fire in the vicinity.

(e)   **Operation of acetylene generators**

There should always be enough water in the generator to generate and cool the gas. Water containers and water seals should always be kept free from impurities, carefully filled and tested. Generators should not be illuminated with open flames. Leaky places should be detected by soaping or by smelling. Frozen generators and hydraulic seals should not be thawed with fire or red-hot iron, but only with hot water or steam, and at a safe distance from fires and stoves.

Before a generator is restarted it is absolutely necessary to see that the safety valve, the water supplies and the connecting pipes have completely thawed.

(f) <u>Repairing generators</u>

If a generator has to be repaired it is best to let the manufacturer do it. If this is not possible, and no specialised workshop is available, the work should be done by a competent person who is familiar with the generator and has been well instructed in the accident risks. Before repairs are begun, the generator should be carefully cleaned and then twice filled to overflowing with water so that all remains of gas have been expelled. This precaution is necessary even if no hot work is to be done. If any hot work, including soldering, is to be done, while it is in progress the generator should be kept full of water, or if this is not possible it should be filled with steam, carbon dioxide or nitrogen and continuously flushed out. If these precautions are not taken there will be a risk of an explosion. Many explosions have caused very serious and even fatal injuries.

1.9.8 <u>Gas cylinders</u>

Cylinders of oxygen, compressed air and fuel gases are classed as transportable closed containers for compressed gases, liquefied gases or gases dissolved under pressure. They are therefore subject to special regulations regarding manufacture, design, equipment and handling.

(a) <u>Testing and identification of gas cylinders</u>

Cylinders authorised for use should conform with the relevant technical principles and have been tested by an official expert. Cylinders should be retested at officially determined intervals, and should be filled only with the gas named on them. The gas should be identified by a colour on the outside of the cylinder as follows:

Oxygen	Blue
Acetylene	Yellow
Hydrogen	Red
Compressed air	Grey

(b) <u>Transport and use</u>

During transport and storage, the protective cap should be on the cylinder and after a cylinder has been used the cap should be immediately screwed on again. Cylinders should be handled carefully, and in particular they should not be thrown or rolled. Cylinders for different gases should not be stored together, nor should any cylinders be stored with highly combustible materials.

Filled cylinders should be protected against heat, especially against prolonged exposure to sunlight, and also against sharp frost. They should not be struck, knocked or jolted. Standing cylinders can be prevented from falling by means of fixed or mobile racks, clamps, chains and the like. They should not be installed even temporarily for use near smithies, stoves, heating appliances, etc. Cylinders may not be laid flat for delivering gas. Care should be taken with acetylene cylinders that the valve is at least 40 cm higher than the foot. During lengthy breaks in the work, both the burner valve and the cylinder valves should be closed, and this applies not only to the close of work but also to meal breaks and shift changeovers.

(c) <u>Cylinder fires</u>

If all proper precautions are taken, it is easy to prevent cylinders from catching fire but every welder should know what to do should a cylinder catch fire. If the fittings of an oxygen cylinder burn the valve should be closed immediately. Cylinders that have been on fire should be marked and taken out of use.

Acetylene-cylinder explosions are the result of decomposition of the acetylene which leads to large rises in temperature and pressure. Rises may be inferred when, after a backfire, the cylinder wall heats up beginning at the top, or the gas streaming out of the valve is sooty or smokey, or has an unusual smell. With cylinders that have been directly heated from the outside by fire or radiation there is always the risk of decomposition of the acetylene. Cylinders in which decomposition has begun should be continuously cooled with large quantities of water from a safe place at a good distance. If the resources of the undertaking are insufficient the fire brigade should be called. The surroundings should be cleared for an exploding cylinder may be flung several hundred metres.

1.9.9 <u>Fittings and welding appliances</u>

(a) <u>Oxygen fittings, piping and packing</u>

Parts of fittings and piping that come into contact with oxygen should be freed from grease, glycerine and oil before the cylinder is used. Pressure gauges for oxygen should bear an inscription worded "Oxygen, keep free from grease". The fittings and packing should also be kept free of grease during use so as to avoid the risk of an explosion. Even slight traces of greasy substances are enough to start a fire or an explosion. For this reason fittings should not be touched with oily cleaning rags or greasy fingers. Splashing or dripping oil or grease may also cause a fire.

(b) <u>Acetylene fittings, piping and packing</u>

Copper should not be used for acetylene fittings, piping or packing, but the use of alloys with up to 65 per cent copper is permissible.

(c) <u>Reducing valves</u>

Reducing valves should be so designed that neither the cylinder nor persons in its vicinity can be endangered by tongues of flame issuing from them. The safety valves of reducing valves should blow vertically upwards. When a reducing valve is being screwed on care should be taken that the springloaded lid is directed vertically upwards or downwards.

Reducing valves for acetylene should not be set for any pressure higher than 1.5 at its gauge. Reducing valves may freeze when large quantities of oxygen are drawn off quickly because the drop from the high pressure in the cylinder to the low working pressure causes a sharp drop in temperature. Freezing of reducing vales for oxygen can be prevented by using electrical heating cartridges.

For all reducing valves there should be a test certificate from an officially recognised testing station. Approved types of valve will bear the station's test mark, as well as the maker's name and the factory number.

(d) Gas hose

Gas hose should be securely fastened by clamps or similar means. Wire is not suitable for fastening hose. Hose should be maintained in good condition and protected against damage from kinks, burns and from being run over, and also against impurities in oil and grease. Even slight damage should be immediately repaired.

Leaky hose can be a cause of fires and explosions if the escaping oxygen or fuel gas penetrates clothing or hollow spaces. Consequently hose should not be led between the legs or over the shoulders.

(e) Burners

Burners should be equipped with separate stop cocks or the like for the fuel gas and the oxygen. In most cases injector burners are used; in these, because of its high pressure, the oxygen pulls the fuel gas with it on the injector principle. Before the burner is used it should be seen to be pulling properly. If a burner repeatedly pops or backfires, it is often caused by overheating or clogging of the mouthpiece, or insufficient tightening of the cap nut by hand. The suction of the injector should be powerful enough to enable the fuel gas and oxygen to mix in the right proportion and the flame to be made large or small.

(f) Handling of burners and fittings

Burners, reducing valves and other fittings should be maintained in good condition and protected against dirt. If possible arrangements should be made for the burners to be safely laid down or hung up during work. They should not be hung on gas cylinders or acetylene generators.

During lengthy interruptions in the work burners and reducing valves should be kept under lock and key in a place where they will not be exposed to dust. Burners and oxygen reducing values should not be stored with oily or greasy objects, or in oily or greasy places. Cases for housing assembled appliances and hose should have sufficiently large ventilation openings. Connected burners should not be placed in closed containers such as tool chests.

1.9.10 Special kinds of welding

(a) Welding on or in containers, etc.

When welding vessels, containers, appliances, piping, pits and the like it is always well to assume that they have contained dangerous substances. Consequently before work begins a competent person should ascertain whether under the influence of heat flammable, explosive, comburent, or noxious gases, vapours or dusts could form. If this cannot be definitely ruled out the containers, etc., should be treated as though they contained dangerous substances. Owing to the seriousness of the risks, the work should be done only by experienced persons on express orders and under competent supervision.

For the rest, the following measures should be taken before hot work is begun. Plugs, stoppers and the like on containers should be opened carefully without using sparking tools or flames. All residues should be repeatedly flushed out with steam or hot water. The containers should be filled with water to just below the heating point, and kept full during the work by means of suitable appliances such as swivelling pipes or hose. If this is impracticable for overriding reasons steam or an inert gas such as nitrogen or carbon dioxide should flow through the container while the work is going on. If the occurrence of harmful gases may be expected and there is no exhaust system, respirators should be worn. See also section 2.4.6 on respiratory protective equipment.

(b) Welding in confined spaces

For welding in confined spaces (tanks, boilers, containers) mechanical ventilation is indispensable because of the risk of nitrous gases that always form, and sometimes also because of the vapours and fumes that form as a result of the action of the welding flame on materials that may be present. Even traces of nitrous gases may cause dangerous illness, that often becomes manifest only several hours afterwards and may end fatally.

If poisoning is suspected the work should be immediately stopped and the victim laid down and kept quiet in the fresh air until the arrival of the doctor. For welding in confined spaces effective ventilation should be ensured by combining a fresh-air supply with local exhaust. If adequate ventilation is impracticable, suitable air-line respirators, as far as possible with a compressed-air supply, should be worn. Hanging up a compressed-air hose in a confined space can never be considered as an adequate means of ventilation.

It is not permissible to ventilate a confined space by blowing in oxygen or using oxygen breathing apparatus because even with a slight increase in the oxygen content of the air working clothes become permeated with oxygen and if a spark reaches them they will immediately burst into flames.

1.9.11 Other precautions

(a) Eye protection and protective clothing

In welding operations suitable eye protection should be provided against sparks, heat and visible and invisible rays. Not all coloured glass affords protection against rays. See also section 4 on personal protective equipment, part 2.4.3 on eye protection.

When the work so requires other means of protection should be worn, for instance aprons, gloves, footguards and protective clothing. Working clothes soiled by flammable substances such as oil, grease, petroleum and benzene should not be worn. When work is done in confined spaces the persons employed in them must wear fire-resistant protective clothing. See also section 1.9.12 on entering containers, etc.

(b) Health protection

Respirators should be worn when lead fumes or lead dust may be evolved during welding of objects containing lead or treated with red lead or lead paint. See also section 4 on illness caused by lead and its compounds and 4.6 on respiratory protective equipment.

(c) Fire protection

One must always reckon with a fire risk in welding, especially if carried on in rooms with a fire or an explosion risk. The danger is not always in the welding flame itself, but sometimes in the heat that it radiates or in splashing or falling incandescent metal. There is also a fire risk in rooms over, under or adjoining the workplace. These risks can only be avoided by advance precautions. These will include not starting welding until after a thorough inspection of the workplace and its vicinity including all adjoining premises, the removal of stocks of highly combustible materials, the provision of fire-resistant coverings for walls and floors, the wetting of neighbouring wooden parts, and the keeping in readiness of water supplies or suitable fire extinguishers. After the work has finished either a fire watchman should remain at the workplace or a scrupulous fire inspection should be carried out at frequent intervals, so that any incipient fires can be detected and extinguished in good time.

(d) Protection of women and young persons

Welding should be done only by reliable persons over 18 years old who are familiar with the equipment and the operations and have been suitably trained. Unqualified persons and persons under 18 years old should only do welding under supervision, and the person supervising should remain permanently at the workplace.

Women should not be employed on welding work involving special risks, including work in confined spaces, work producing excessive heat, and welding objects that are galvanised, contain lead, or are coated with paint containing lead.

Entering containers and pits

1.9.12 General

Serious accidents, most of them fatal, repeatedly occur because persons enter containers, pits, sewers, wells, etc., without ascertaining whether toxic or narcotic gases have accumulated in these confined spaces, and if they have, without the necessary precautions being taken. The frequency and severity of these accidents clearly show how important and necessary it is to recognise the dangers and to know the precautions required. Dangers in containers and other confined spaces may be caused as follows:

- harmful gases or vapours in dangerous concentrations. They may arise from known substances that were kept in the container, but also from sludge residues, or residues adhering to the walls, or to gas welding and cutting operations, from handling or spraying paints or insulating materials or from substances entering sewers;

- oxygen deficiency, which can lead to asphyxiation. This can be caused by a protective gas used to expel an air-oxygen mixture and so reduce an explosion risk, or by substances that absorb or transform the oxygen in a container. It can also be caused by unsuitable and insufficient ventilation during work in a confined space.

- corrosive or toxic substances that may be touched or inhaled;

- flammable gases or vapours that may cause fires, conflagrations or explosions.

Any of these risks may also occur in certain circumstances near container openings.

1.9.13 Precautions

Precautions to be taken against the dangers threatening persons who enter confined spaces include (a) the issue of a written entry permit, (b) cleaning, (c) respiratory protection, (d) explosion prevention, (e) protection against excessive contact voltages, (f) protective clothing and (g) rescue arrangements.

(a) **Entry permit**

Containers and other confined spaces in which dangerous gases are present or may accumulate, or in which an oxygen deficiency may occur, should be entered only subject to a written permit and on the responsibility of the manager or his representative authorised for the purpose. A special permit and the application of suitable precautions may also be necessary for the performance of certain work in the vicinity of container openings, or when lamps or other appliances are taken into a container in which there are gases or vapours that could be ignited by the appliances.

Special forms of entry permit should be provided with a text adapted to the requirements of the undertaking and the conditions in question. A permit should not be issued until the responsible person has satisfied himself that the atmosphere in the container or confined space is safe and that the necessary precautions have been taken. Testing the composition of the atmosphere with a bare flame (which would reveal an oxygen deficiency if it went out) is insufficient and also dangerous because of the possibility of an explosion. It should therefore not be attempted.

When devising precautions it should be remembered that heavy gases and vapours may accumulate at the bottom of a container or other confined space, and that stirring residues, scooping up sludge, removing deposits, knocking off rust and similar operations may release dangerous gases or enable them to form.

The responsible person should supervise the application of the precautions required before a container or other confined space is entered.

(b) Cleaning

Before it is entered a container or other confined space should be cleaned as thoroughly as possible, for example by repeatedly filling it to overflowing with water or by spraying or flushing it with large quantities of water or other suitable liquids while stirring any sludge-like residues. Cleaning with water will be successful only with soluble residues. In some cases spaces can be cleaned by steaming them out.

After cleaning it should be ascertained whether the atmosphere in the space still contains any dangerous gases, and if necessary cleaning should be repeated.

If dangerous quantities of harmful gases are liberated during cleaning operations the cleaners should wear suitable respirators, and if appropriate, protective clothing. With flammable gases and vapours the fire and explosion risk should be taken into account.

(c) Respiratory protection

Containers and other confined spaces in which toxic or narcotic gases and vapours may accumulate should only be entered by persons wearing respiratory protective equipment that functions independently of the dangerous atmosphere (air-line respirators, oxygen breathing apparatus, compressed-air respirators) unless cleaning as described in section 1.9.13(b) is possible. If, while persons are inside, a container or other confined space is adequately ventilated by blowing fresh air in or through so as to protect the entire working zone, and if it is ensured that the concentration of harmful gases and dusts does not exceed the permissible maximum, respiratory protective equipment need not be worn provided that a careful examination has been made and the responsible person agrees that the equipment is not required. Owing to the fire risk oxygen should not be used for ventilation.

(d) Protection against explosions

If flammable gases or vapours can accumulate in dangerous quantities inside a container or other confined space it must be treated as having an explosion risk. Ventilation should ensure that the lower explosive limit is not reached. Lamps should be prevented from falling and protected against other damage: damage to electric conductors should be prevented. Suitable fire extinguishers should be kept in readiness.

(e) Protection against contact voltages

If electric lamps and appliances are used in containers with highly conductive walls precautions against contact voltages will be necessary. For hand lamps only low voltages up to 42 V should be used. Before use lamps and their leads should be carefully examined to ensure that they are in perfect condition.

(f) Protective clothing

If persons have to enter containers or other confined spaces in which there may be remains of corrosive or other harmful substances, they should wear suitable protective clothing.

(g) Rescue arrangements

Only persons wearing a safety belt should enter a container or other confined space for the purpose of rescuing someone who is sick or injured. The rescuer should be constantly watched by a strong and reliable person outside the space. The rope should be kept taut and securely fastened outside. It is useful to have a safety belt with shoulder and leg straps and with the ring for fastening the rope placed at the nape of the neck.

It should be possible for the watcher to get help without leaving his post. Rescue operations should only be undertaken when assistance has arrived. All the rescuers should be roped and if necessary equipped with respiratory protective appliances. This may also apply to the watchers and supervisors, depending on the circumstances.

1.9.16 Special precautions for repairs, hot work and painting

If hot work such as gas welding and cutting and soldering is to be done in or on the outside of a container or other confined space an additional written permit should be obtained from the responsible person. Fuel gas cylinders and acetylene generators should not be taken into confined spaces. Welding and soldering equipment should be taken outside during any interruptions of the work, including the usual breaks. For the rest the recommendations in section 1.9 on gas welding and cutting, and especially section 1.9.10 on special kinds of welding should be followed.

When laying on or spraying paints and insulating materials containing solvents the work should always progress from the bottom upwards. Fresh air should be continuously supplied from above and there should be an effective exhaust at the lowest point. The harmful gases and vapours should be so exhausted that they do not have to pass by the workers. If this cannot be satisfactorily arranged, respiratory protective equipment should be worn.

Lifting and carrying loads: loading, shifting and stacking objects

1.9.17 General

It is not always equipment and methods of working that are responsible for accidents. Serious accidents often occur in the simplest operations such as lifting and carrying loads, or loading, shifting and stacking objects, either because the most obvious precautions are not taken, or because the workers are inexperienced and work incorrectly. Usually the workers themselves are not to blame, especially if the supervisors have failed to instruct them in proper methods of working. Such instruction is urgently necessary, if only because most of the workers concerned are unskilled: often it is not even ascertained whether they are physically fit for the work.

1.9.18 Lifting and carrying loads

Lifting and carrying loads, that is manual transport, is mostly heavy work that causes intense fatigue, and hence, not infrequently, accidents. The work imposes not only a permanent strain, but sometimes also peak loads on the back and the heart. These loads are an important factor in occupational injuries of the posture-control and locomotor systems and in the wear and tear of the heart. The term "heavy work" has only relative significance when applied to lifting and carrying loads for individual working capacity varies widely. Work that is easy for a strong young man, might be heavy, and even unbearable strain for a weak or an old man, and even more a woman or an adolescent.

Recent research has shown that man's spinal column is an imperfect instrument for lifting and carrying heavy loads because of its construction and the changes undergone in ageing; if it is overworked disc injuries and excessive wear may result.

(a) Measures for reducing disc injuries

So far as possible lifting and carrying of loads should be mechanised by use of lift trucks, lorries, block and tackle, lifting appliances, etc. If mechanisation is impossible and persons of poor physique are employed every effort should be made to reduce the loads. The maximum permissible loads prescribed for lifting and carrying by women and young persons vary in different countries.

The choice of suitable workers is always important: if loads are heavy only strong, healthy men should be employed. Even so, this is not enough: they should also be fully instructed in the handling of loads.

(b) Lifting and carrying techniques, lifting or picking up loads

It is first necessary to see whether the load can move freely. If large unwieldy loads are to be taken on the back it is best to get help. The loads should not be lifted jerkily, or with a bent back, but with bent knees.

Carrying loads

As far as possible loads should be evenly distributed over the body, preferably by means of appliances such as belts, racks or yokes. Heavy and bulky loads should not be carried in front of the body but on the shoulders or the back. On long journeys and on inclines and stairs loads should if possible be set down frequently at carrying height.

Setting down or throwing down loads

Loads should be set down evenly, and care should be taken that they are not caught just before they are set down. During setting down the knees should be well bent and the back kept straight. If the load can be thrown down it should be ensured that no one will be struck by it.

Carrying of loads by a number of persons

There should be enough carriers so that if one falls out the remainder will not be overloaded. They should be so placed that they do not get in each other's way and arranged in order of size; only one should give orders. When the appropriate order is given the load should be lifted, lowered or thrown by all carriers simultaneously.

Other recommendations

For protection against objects with sharp edges, hand leathers or shoulder pads may be worn. To avoid crushing fingers and feet, landing places should be provided with supports on which the load can be set. When heavy loads are being set down foot injuries can be prevented by safety boots. Persons carrying open containers holding dangerous substances should wear safety goggles.

1.9.19 Loading, moving and stacking
objects

Objects should always be so loaded, unloaded, moved and stacked that persons cannot be endangered by objects falling, rolling, overturning, coming to pieces or breaking.

(a) Loading and moving

Before wagons are loaded or unloaded precautions should be taken to ensure that they cannot inadvertently be set in motion during the work. When short heavy objects such as casks are being loaded or unloaded by hand, dray ladders or the like should be used and secured against displacement.

When heavy loads are being moved on inclines no one should be below the load. To control the movement of the load, in addition to the necessary wedges and sprags, ropes or other means of haulage should be used.

If rollers are used to move heavy loads the direction of moving loads should not be changed with the hands or the feet but with poles or hammers.

Beams, round timber, steel girders and the like should not be thrown down when they are being unloaded from wagons by hand.

(b) Stacking objects

Goods should be so stacked that they do not interfere with the operation of machines, appliances and plant, or hinder the use of passageways and other trafficways.

Goods should be stored on firm supports that will not subside, and their weight should be limited so that the ground will not be overloaded. Freshly concreted floors should not be loaded until the concrete has completely set.

On scaffolds, as with all building parts, care should be taken that building materials are not stacked on the floor in excess of the maximum permissible load.

When they are stacked objects and materials should not exert any lateral pressure on partitions unless they are strong enough to withstand it. Stacks should not be carried to a height at which their stability would be impaired. In particular the following precautions should be taken.

When filled sacks are stacked the mouths should be turned inwards. At stack corners in the open the first four end bags should be cross tied, and a step back of one bag should be made at every fifth bag in height.

In order to obtain the necessary stability sawn lumber should be laid on supports above ground level with horizontal or slightly inclined layers separated by tie pieces, the ends of which should not project into passageways.

Pipes and bars should be placed on stable storage racks so arranged that the material can be safely withdrawn. If racks are not provided the pipes or bars should be stacked in layers resting on wood strips with stop blocks at the ends.

2. **Equipment**

The safe condition of equipment is one of the most important prerequisites for the prevention of occupational accidents and diseases. The requirements necessary to secure this end may be summarised in the following general principles:

(a) the design and construction of equipment should comply with technical safety rules. If the design of equipment does not exclude all risk of accident, it should be provided with the appropriate safety devices;

(b) equipment should only be used for the purpose for which it was designed. The manufacturer's operating instructions, and any instructions for erection and dismantling, should be strictly followed;

(c) persons entrusted with the erection, operation or handling of equipment should have been trained for the purpose, and have shown that they possess the necessary knowledge and experience;

(d) equipment should be looked after carefully, properly maintained, thoroughly examined for defects before use, and properly serviced and used. Defective equipment should not be used;

(e) if the regulations applying to the equipment provide for tests on delivery or at intervals or otherwise, the tests should be carried out at the prescribed times.

Experience has shown that general instructions are not enough to secure the application of these principles: the management, or supervisors appointed for the purpose, should always exercise control of their observance.

2.1. **Machinery**

In this section when reference is made to general safety requirements for power transmission and working parts they should be taken as applying not only to equipment that can be considered as machines in the strict sense of the term, but also to all other equipment that has such parts, including, for example, equipment for the vertical or horizontal transport of building materials and parts that belong to the group of lifting and transport appliances.

Building machinery in the strict sense comprises:

machines for working stone, wood or metal;

mixers for mortar or concrete;

compactors for compacting earth or concrete;

pile drivers;

cleaning machines.

Building machinery is usually driven individually by internal-combustion engines or electric motors. Although this is not always recognised the individual drive is itself an important contribution to accident prevention, because it replaces transmission machinery which is particularly dangerous. The significance of the individual drive as a factor in accident prevention can be inferred from the number and variety of safety measures required for power transmissions, as indicated below:

(1) enclosure or guarding of all moving parts in the working area, for instance, flywheels, shafts, gear wheels, belts, ropes and pulleys;

(2) guarding of belts and ropes if they pass over workplaces or pass through traffic zones;

(3) the provision of mutually independent stopping devices on working machines that are driven from a common prime mover;

(4) the protection of these stopping devices against inadvertent actuation;

(5) prohibition of throwing belts on and off by hand or repairing belts while the machine is running;

(6) prohibition of stepping over moving belts;

(7) the use of special appliances for laying on, shifting or throwing off moving belts.

In addition to the guarding of toothed and chain drives of all moving parts and the in-running, nip and shearing points in working or traffic areas, machine protection entails a series of other generally applicable safety measures as set out below:

(1) observance of technical rules for machine design ensuring operating and productive efficiency;

(2) orderly arrangement of controls clearly visible from an easily accessible and safe operator's stand;

(3) competent working and maintenance of machines, including regular inspection of their condition;

(4) operation by reliable workers whose competence has been established;

(5) prohibition of machine operation by workers unless they have been explicitly authorised.

In addition, special precautions are necessary according to the type and purpose of the machine; these will be referred to at the appropriate place.

As can be seen from these principles the main responsibility for machine protection falls on the manufacturer. He can only be considered to have performed his task properly when the man working at the machine is safe in all conditions and in all situations. This will not be achieved by safety devices alone. At times the safety device is an addition that the operator finds a hindrance or a nuisance, and if it can be removed it may not be used. As far as possible moving machine parts should be enclosed and, instead of being provided with additional safety devices, the machine and its tools should be safe by design.

However, if safety devices are necessary they should if practicable be automatic and independent of the user's volition. Danger points that cannot be enclosed should be protected by automatic stopping or disengaging devices.

On machines where not only the drive but also the different controls are electrically operated, the operation of certain safety devices can be incorporated in the arrangements; their efficacy as safeguards is considerably enhanced by their automatic action. On all electric machines uninsulated live parts should be protected against accidental contact.

Working and traffic zones in which protection for machines is necessary are calculated on the basis of a man's reach and extend 2.50 m upwards and 0.70 horizontally from his stand, or from enclosures that are not safe barriers. It should, however, be realised that these zones may be larger, for there may be danger points at greater distances if long objects are regularly used.

In addition to the safe design and equipment of the machine, proper operation and servicing are also required. Consequently, the contractor must have, or train, qualified machine operators. As a rule machines should only be operated by persons of at least 18 years of age who have been found suitable and to have experience of the operation of the machine in question. Safety devices, guards, enclosures, etc. protecting dangerous machines or machine parts should not be removed or put out of action by the operator, except when the machine is stopped and for the purpose of immediate repair or adjustment. As soon as the repair or adjustment is finished the protection must be replaced. If defects are found in machines or safety devices they should be immediately reported by the operator or any other worker who notices them. The machine should be stopped, the starting mechanism blocked or a conspicuous notice affixed to the machine prohibiting its use until the necessary repairs have been carried out and it is once again in proper working order. Competent operation of a machine implies proper care and regular maintenance. The governing principle here is that repair and cleaning operations on machines may only be carried out when they are stopped, and measures have been taken to prevent them from being inadvertently started up.

Lubrication of machines by hand should not be allowed unless it can be done safely. Bearings that cannot be reached safely should be provided with long oil or grease supply pipes.

Prime movers

Internal-combustion engines

With internal-combustion engines there is a considerable risk of accidents from the kick-back of the handle following a backfire when starting up the engine by hand. The starting device should therefore be so designed that it is automatically disengaged when the engine starts up and cannot kick back. Kick-backs can also be prevented without this precaution if ignition is automatically delayed when the engine is being cranked. With diesel engines it is sufficient to have a cranking device that is automatically disengaged when the engine starts up.

However, when internal-combustion and diesel engines are being cranked accidents may be caused merely because the handle falls out or is thrown out of inadequate seating. This sort of thing can be avoided by double bearings.

Another accident risk with internal-combustion engines lies in the generation of poisonous carbon monoxide when an engine is run in unventilated or poorly ventilated premises (see the section on dangerous gases and materials).

For protection against noise see under 10.3.4(a).

Electric motors

In the use of electric motors for driving machines consideration should be given to the measures described in section 2.2, electrical equipment and appliances. This also applies to electric control appliances for machines.

Working machines

2.1.1 Circular saws at building operations and portable woodworking machines

(a) **Circular saws at building operations**
ILO code of practice, safety and health in building and civil engineering work, section 14.2

The circular saw is always one of the most dangerous woodworking machines, but its many uses on building sites are accompanied by additional risks. Not every circular saw is suitable for building sites; their special requirements must be met by design and equipment: the skill and reliability of the user is also important.

The chief risks with circular saws are (a) that the workpiece will kick back when at the end of the cut it jams on the back of the blade and is thrown upwards: according to the violence of the kickback the worker's hand may slip and hit the saw blade or the wood will be thrown back and (b) that when guiding the workpiece the worker's hand will come into contact with the blade.

These risks can be largely eliminated by good design. In particular the following conditions should be satisfied.

Saw bench: to enable workpieces to be properly guided, saw benches should not be too small, if only for the reason that in building operations circular saws not infrequently have to deal with fairly large pieces, which are very difficult to guide on small benches. For a blade diameter of 450 mm the bench should measure at least 660 x 1,000 mm, and for a larger diameter, at least 850 x 1,250 mm.

The bench should be smooth, even and free from cracks. The slot for the blade should be quite straight and on metal benches should be edged with wood.

Saw blade: the choice of the right blade is important from the safety standpoint. For building operations the wolf's teeth type is particularly suitable. The blade should be sharp and perfectly set, and should be free from burns and cracks, or it will wobble.

Riving knife: to avoid dangerous kick-backs a riving knife should be fitted, but it will only serve its purpose if it meets the following requirements as regards design, mounting and adjustment (Fig. 6).

It should be so thick that it does not bend: it must be thicker than the saw blade but not thicker than the cut. It should be so mounted on a special support that it can be quickly and properly adjusted horizontally and vertically in the plane of the blade. It should be secured against inadvertent displacement, twisting and flying out. It should be not more than 3 mm from the edge of the blade and should be so adjusted that its tip is not below the root of the highest tooth. If blades of different thicknesses are used on a circular saw, riving knives of different thicknesses must be available.

Hood guard: to prevent the danger of contact with the blade, circular saws should be equipped with a hood guard that covers the blade to the cutting zone and should be adjustable. On saws with a blade diameter up to 450 mm the hood may be fastened to the riving knife. With larger diameters the hood can conveniently be mounted on a special support fitted to the bench or mounted in some other suitable manner (Fig. 8).

Enclosing the blade under the bench: the blade should be so enclosed under the bench that the enclosure extends about 10 cm beyond the teeth. It should be so fitted that the blade can easily be changed, and so designed that the sawdust can easily run off without clogging.

Adjustment of height of cut: especially valuable in accident prevention is a device that enables the cutting height of the blade to be limited to the thickness of the workpiece. Proper adjustment reduces the part of the blade above the bench to the minimum required for cutting so that accidental contact with the teeth is largely prevented.

Accessories: the variety of jobs to be done on a circular saw in building operations make it necessary to provide special safety devices. These include a guide for cross-cutting, a guide for rip-sawing and a holder for diagonal cuts such as wedges.

The guide for cross-cuts should be easily adjustable, rust-proof and weather-proof. It should be so made that it cannot tilt as the workpiece advances.

The guide for rip-sawing can be mounted on a tubular fitting or in a groove: it should be firmly affixed without any play.

When used in conjunction with the guide for rip-sawing the wedge holder enables wedges to be cut safely. It must be so made that the guiding hand does not come dangerously close to the saw blade. For pressing the wedge against the holder and for removing the wedge safely after sawing a push stick should be used (Fig. 7).

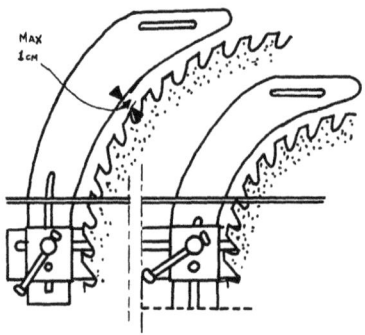

Fig. 6
Adjustable riving knife

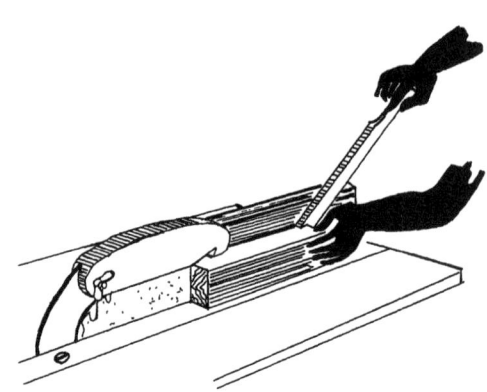

Fig. 7
Use of push stick

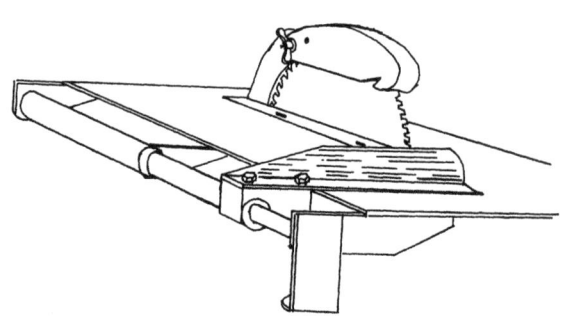

Fig. 8
Circular saw with blade guard

Notice: circular saws should be operated only by persons with the necessary skill and experience and a plate should be fitted to every saw prohibiting its use by unauthorised persons.

(b) **Portable woodworking machines**

Portable woodworking machines owe their development mainly to the needs of building sites for framing (for concrete and reinforced concrete structures) and for carpentry. These machines include circular saws, band saws, rotary planes, chain mortising machines and drilling machines.

While the particular danger of fixed woodworking machines, which lies in the manual feed, is absent in portable machines because they are moved up to the work, on the whole portable machines are just as dangerous as fixed. As regards accidents with portable machines a distinction can be drawn between those due to the manual operation of the machine, those due to electricity and those of the same type that happen on fixed machines.

Accidents in the use of portable machines can be mainly ascribed to faulty operation, as when the machine is not switched off after each cut or when it is moved to another workplace before the blade had stopped. Accidents also occur when the machine is momentarily laid down without being stopped, because in such cases the rotating tool will cause the machine to move.

Accidents at drilling machines occur when they are operated without a stand. For instance, if the drill jams in the work the machine may rotate and the operator will lose control of it. If this happens when he is working at a height he may be thrown down.

Electrical accidents caused by dangerous contact voltages are particularly numerous.

Sometimes portable machines are used as fixed machines without the conditions necessary for safe feeding and guiding the work. The machine tables are frequently much too small and makeshift; the protective devices required for fixed operation are usually lacking.

Portable circular saws

According to their size, which depends on the depth of the cut to be made, portable circular saws are operated by one man or two men. The smaller types have a fixed protective hood, and the larger a fixed upper hood and a movable lower hood. The protective hood does not give complete protection, because the blade has to be exposed for sawing. On the smaller types with a fixed hood, disengagement of the cut-depth control after the cut causes powerful springs to return the blade to the idle position, under the fixed hood.

On portable circular saws with a movable lower hood, this hood slides into the upper hood to an extent varying with the size of the workpiece and the depth of the cut, and so exposes the blade during the cut. When the cut is finished and the machine raised, the movable hood swings back to its idle position and again completely covers the blade. It will thus be seen that the efficiency of this guard and its value as a safety device are limited. Consequently the safety requirements for portable circular saws are as much a matter for the manufacturer as for the user:

- the design of the protective hood must ensure its mobility in diagonal as well as straight cutting. However, in the event of a stoppage a special knob should enable the hood to be pushed back safely by hand;

- the design of the hood should afford good visibility of the line of the cut and also allow unobstructed escape of sawdust;

- the control switch should be in the immediate vicinity of the handle so that it can be used without a change of grip;

- smooth movement of the movable hood should be ensured by proper maintenance;

- workpieces should be so positioned that they cannot shift;

- the cutting depth of the saw blade should be adjusted so as not to exceed that required for the cut;

- the motor should be switched off and the blade stopped after each cut, when clearing stoppages, and when changing workplaces;

- the saw should not be laid down until it has stopped and the blade put in the neutral position.

Portable band saws

There are two types of portable band saws - for one man and for two men. The two-man type is preferable. It is particularly important to place the electric cable properly. If the worker changes his place while working the cable may be caught on the running blade. To prevent this there should be an attachment on the machine frame that will hold the cable even if the operator releases his grip.

Portable drilling machines

With these machines a typical accident cause is jamming of the drill in the wood so that the machine is whirled around, which can be particularly serious at elevated workplaces. Consequently drilling should not be done unless a clamping device is used.

Portable woodworking machines used as fixed machines

Besides being used in the manner for which they were designed, portable circular saws and rotary planes are often used as fixed machines. For such use the machines must be equipped with the same safety devices as are required for fixed machines of the same type, namely, riving knife, hood guard, guides, etc. for circular saws, and covering of the cutter block on planes used as overhand planing machines.

One of the chief causes of accidents when portable woodworking machines are used as fixed machines is the table, which is usually makeshift or too small, so that large workpieces cannot be properly fed in or guided. Firm and sufficiently large tables are an absolute necessity if portable machines are used as fixed machines, even if only for a short time. Rotary planes need a suitable cover for the cutter block to make it safe, for otherwise the deep grooves in the block used to remove shavings may cause severe hand or finger injuries.

For protection against noise, see 10.3.4(f).

2.1.2 **Machines for bending and cutting steel reinforcement for concrete**

(a) **General**

As electrically-powered machines, machines for bending and cutting steel reinforcement for concrete require an electrical installation conforming to the safety regulations that will prevent dangerous contact voltages on the machine housing.

See section 2.2, electrical installations.

(b) **Machines for bending steel reinforcement for concrete**

When these bending machines were equipped with a bending flap there was a danger that while the machine was being operated there would be a nip point between the edge of the machine table and the flap as it projected in its movement over the table. These nip points were the source of serious accidents, some of them fatal. To avoid them it was necessary to fit a protective hoop suited to the area of movement of the flap and preventing the occurrence of a nip point. This precaution, however, proved insufficient because when adjusting the bending tools the machine operator often found the hoop inconvenient or a nuisance, so he removed it for adjustments and omitted to replace it when starting up the machine. Suitably widening the table proved more effective because it served the same purpose as the hoop, and the protection it afforded could not be rendered ineffective.

Recent developments in bending machines have resulted in types in which the bending process is effected by plates. These machines and their tools are so designed as to prevent the occurrence of dangerous nip points.

(c) **Machines for cutting steel reinforcement for concrete**

To avoid finger injuries on power-driven shears the wall of the housing against the mobile knife should be set back on one side. A special roller arrangement should make it possible to draw round steel bars out of the shearing opening.

2.1.3 Stone-cutting and stone-grinding machines

(a) General

However varied the functions of stone-cutting and stone-grinding machines they are all essentially of the same type. At a power-driven cutting or grinding tool natural or synthetic stone or concrete is split, cut or ground. The only differences are in the form and composition of the cutting or grinding tools.

Apart from the risks of the mechanical drive, among typical risks of stone-cutting and grinding machines are bursting of the cutting or grinding tool and, if siliceous material is worked, inhalation of the harmful dust raised. Cutting or grinding tools may burst for several reasons - faulty treatment of the tool, faulty storage, improper mounting on the driving shaft, but chiefly, exceeding the prescribed maximum peripheral speed of the cutting or grinding tool.

(b) Cutting and grinding wheels

Particulars of abrasive wheels

All abrasive wheels should bear an inscription giving the following particulars:

manufacturer's name or registered trade mark;

dimensions when new;

maximum permissible number of revolutions per minute when new, and corresponding peripheral speed in metres per second;

nature of bonding;

test number (authorisation mark) and special instructions for use if available or necessary.

For every abrasive wheel of 100 mm or more diameter these particulars, clearly legible, should be given either on the wheel itself or on an adhesive label. The wording should be in the language of the country where the wheel is being used. The manufacturer's name or mark should be so placed that it is legible so long as the wheel is in use.

Abrasive wheels with a peripheral speed of over 35 m/s and wheels with magnesite bonding should also be conspicuously marked by a coloured stripe running diagonally across the adhesive label and across one side of the wheel. Identification colours are as follows:

blue for peripheral speeds of 45 and 50 m/s;

yellow for a peripheral speed of 60 m/s;

red for a peripheral speed of 80 m/s;

green for a peripheral speed of 100 m/s;

white for wheels with magnesite bonding.

Testing at the manufacturer's

Before delivery abrasive wheels should be tested by the manufacturer or other competent testing body.

Transport, handling, storage

Because of their brittleness abrasive wheels should not be exposed to blows and vibration in transport and handling.

Storerooms should be dry, kept at a fairly even temperature and frost free.

Cutting wheels should be laid flat on an even base; this also applies to thin ring wheels and cup wheels.

Abrasive wheels with mineral and synthetic resin bonding should not be stored for more than two years.

For lifting and transporting heavy abrasive wheels, lifting appliances and suitable transport equipment should be used.

The faces of large synthetic wheels with magnesite bonding should be given a protective coating to prevent the penetration of moisture.

Mounting

Unless the nature of the work or of the wheel requires some other method of mounting, abrasive wheels should be mounted between flanges of steel, cast steel, spheroidal-graphite cast iron or material with similar strength and elasticity.

The wheel should slip easily and without forcing on to the spindle or centering lug of the mounting, but the whole should not be more than 0.5 mm larger than the diameter of the spindle or lug.

Between the flanges and the wheel should be inserted elastic compressible washers from 0.3 to 1 mm thick according to the extent and roughness of the contact surface. For wheels up to 230 mm in diameter with textile reinforcement and a convex centre no washers are necessary. For mounting large synthetic wheels with mineral bonding, rubber washers up to 5 mm thick should be used.

Grinding and polishing wheels with conical holes should not be mounted by forcing or screwing them on conical spindle ends.

Flanges

Flanges should be made of steel, cast steel, spheroidal-graphite cast iron or other material of similar strength, and they should be rigid so that they press uniformly on the wheel when the fastening screws or nuts are tightened up. This applies in particular when protective flanges are used with conical wheels.

Their contact surfaces on the wheel should have the same internal and external diameters, and they should extend at least 1.5 mm beyond the inner surface of the fastening flange.

Minimum diameters of fastening flanges should be as follows:

(1) one-third of the wheel diameter for straight-sided abrasive wheels with a small hole when protective hoods are used;

(2) one-fifth of the diameter of cutting wheels when the peripheral speed does not exceed 60 m/sec;

(3) two-thirds of the diameter of flat wheels that are run with protective flanges instead of a protective hood;

(4) one-half of the diameter of conical wheels up to 300 mm diameter when both sides are in a protective mounting; for larger wheels the protective flanges should be so dimensioned that the wheel projects not more than 75 mm beyond their rims;

(5) 40 mm for abrasive and cutting wheels with textile reinforcement and convex centre and not more than 230 mm in diamter.

For wheels with a small hole the width of the annular contact surface should be about one-sixth the diameter of the fastening flange.

For wheels with a larger hole, one-sixth of the radius (less the radius of the hole) should be covered by the flange, the width of the contact surface being from 3 to 8 mm so that the recessed area is less than the covered area. The inner flange should be mounted rigidly on the spindle so that it cannot twist.

These requirements concerning flange diameter and contact surface also apply to the mounting of recessed wheels. The recessed diameter should be sufficiently larger than the flange diameter to prevent the flange from pressing on the edge radius of the recess.

Protective hoods, protective mounting

Grinding machines with abrasive wheels running at peripheral speeds exceeding 15 m/sec should be equipped with well-secured protective hoods, protective rings or screens of strong material that will hold all the fragments if a wheel bursts. Protective hoods and rings should leave exposed only the part of the wheel required for the work.

On stationary grinding machines, except wheels for internal grinding it should be ensured by suitable arrangement of the wheel and the surrounding machine parts, that even if a worn-down wheel bursts no fragments will fly out. If this is not possible the protective covers should be adjustable.

Protective hoods on portable grinding and cutting machines need not be adjustable. On these machines the side of the wheel away from the machine needs no cover. The wall thicknesses of protective hoods of high-grade deep drawn plate for portable machines running at not more than 50 m/sec peripheral speed should be as in the following table:

Wheel thickness up to:	Wheel diameter			
	51 - 125 mm		150 - 200 mm	
	A	B	A	B
50 mm	2	2	3	3

A = wall thickness of the peripheral part.
B = wall thickness of the face.

When wheels with a smaller recessed depth are used on portable grinding machines with a peripheral speed of 50 m/sec or more they should be equipped with protective rings, which need not be adjustable. Protective rings for deeply recessed wheels (cup wheels) should be adjustable axially.

The minimum wall thicknesses required for protective hoods of stationary cutting machines for peripheral speeds up to 80 m/sec are given in the next table (in mm).

Material	Maximum thickness of cutting wheel	Cutting wheel diameter in mm							
		250-300		301-400		401-600		601-900	
		A	B	A	B	A	B	A	B
Steel plate	1/50 external diameter	4	2	4	3	6	3	8	4

A = wall thickness of the peripheral part.
B = wall thickness of the face.

Instead of protective hoods, protective mounting may be used if the work cannot be done by any other means. Protective mounting at least prevents fragments of the part of the wheel held in the mounting from flying out if the wheel bursts. The following is an example of protective mounting:

Holding flanges two-thirds the diameter of the wheel with safety washers tested by a recognised independent testing institute and authorised by the competent labour-protection authority. This mounting can only be used on plain straight wheels not more than 40 mm thick and intended for peripheral speeds up to 35 m/sec. The wheels should be carefully freed from any adhering fragments of cardboard and glue from washers.

(c) <u>Grinding machines</u>

<u>Construction</u>

On every grinding machine a plate should be affixed giving the number of revolutions of the spindle and the maximum permissible wheel diameter.

The maximum peripheral speed indicated on the wheel may be exceeded by 10 per cent on portable hand-held machines, and by 5 per cent on all other grinding machines when they are running idle.

Grinding machines on which the number of revolutions can be varied should be equipped with an automatic interlock that prevents operation at a higher speed than the machine was designed for. If an automatic speed governor cannot be built in, for example in portable grinding machines with a flexible shaft, a clear and durable notice should be affixed to the machine giving the maximum permissible wheel diameters for the various speed ranges.

Portable cutting machines for making long straight cuts should be equipped with guiding arrangements to prevent the machine from tilting. This also applies to special hand-held cutting machines for cutting lengths of pipes, angle iron and the like.

Use of safety devices

The safety devices should be constantly maintained in good order and continuously adjusted as abrasive wheels wear down. They should be so adjusted as to afford the maximum protection at all times.

Eye protection

For the prevention of eye injuries suitable protective devices such as goggles and screens should be provided and used. Protective screens mounted on the machine should be large enough and adjustable. They should be made of non-splintering glass which can be protected by fitting ordinary window glass on the abrasive wheel side in such a way that it can be easily changed. (See also section 4.2.)

Dust exhaust

With dry grinding the grinding dust should be exhausted at the point of origin. When grinding with synthetically bonded wheels about 98 per cent of grinding dust comes from the workpiece and risk of injury to health depends to a significant extent on what is ground. The exhausted dust should be effectively removed, and should not be just blown away. See also section 8.

2.1.4 Concrete and mortar mixers

According to the design of concrete and mortar-mixing machines a distinction can be drawn between free-fall and automatic machines. In free-fall machines the mixing process is effected by the rotation of the drum, and according to the type of rotation the mixer may be called a continuous, reversing or dumping-drum machine. In automatic mixers a stirring mechanism moves in a fixed trough.

As a rule mixers are equipped with a feed mechanism which supplies the mixing drum or the mixing trough with the ingredients for the concrete or the mortar.

Accidents, usually serious, are caused both by the mixing and the feeding mechanisms. They are due partly to faulty design or equipment and partly to faulty operation of the machines. Persons may be caught by the rotating drum or by the tools moving in the mixing trough. Inadvertent starting up of the drum or the stirring mechanism while the machine is being cleaned is particularly dangerous. Another source of accidents is the unexpected fall or lowering of the charging skip. Mention may also be made of the accidents commonly occurring on all machines and caused by shafts, gear wheels, etc. They have been discussed in section 1.1. For protection against noise see section 10.3.4(c).

(a) **Construction and equipment**

Technical characteristics and identification

Concrete and mortar mixers should be designed and built in accordance with generally recognised technical rules so that all their parts are strong enough to withstand the stresses of operation and to ensure safe operating conditions. Winches of feeding mechanisms should satisfy the technical requirements for winches described in section 1.3.1.

Mixers should carry a durable and clearly legible inscription giving the manufacturer's name, the type of machine, the year of manufacture and the factory number. Winches of feeding mechanisms should also carry the plate required for all winches indicating the lifting capacity, specifications of the ropes to be used, etc. (2.3.1(a)).

Operator's stand, means of access

The controls of a mixing machine should be so arranged that they can be safely worked from a stand. This stand should also afford a view of the feeding mechanism, and the operator should not be endangered by the rope of the charging winch or by the charging skip. If the operator's stand is more than 60 cm above the ground it should be reached by a ladder or by steps. Like all working platforms, elevated operator's stands should be protected by standard railings with two rails and toeboard. Precautions against falls from elevated stands are also necessary when a mixing machine is installed on piling to supply a building from above. The necessary precautions should be taken by the building contractors in accordance with conditions on the spot.

In the operation of a mixing machine, it may be necessary for the operator to climb on it to inspect it or deal with minor troubles while it is running because the flow of concrete cannot be interrupted. For this reason it is desirable to have devices that enable a person to climb the machine without being endangered by moving parts.

Feeding machines

The winch of the feeding mechanism should satisfy the general requirements applicable to winches. The rails for the charging skip should be so made and installed that the gauge cannot alter. It is particularly important to ensure this when rails have to be extended after a machine has been raised. When the charging skip is being raised by the winch even slight variations in the gauge cause jamming, and hence additional friction, which places additional stress on the winch and all its parts. This may lead, among other things, to premature weakening of the winch rope: if this breaks it may be very dangerous for the operator, especially if, as often happens, part of the rope whips out. It is essential to keep the winch rope under constant observation.

When a concrete mixer is raised there should be not only a longer rail track, but also a longer winch rope, but this can only be provided if the dimensions of the winch drum allow. If they do not, a rope wound in several layers on the drum will project over its rims: if it falls over the edge it will not only prevent the feeding mechanism from working smoothly but will itself suffer damage.

The raised charging skip is always a source of danger, and it must therefore be secured against falling by some device. This applies to both the transport and the operation of mixers. Merely blocking the skip by means of the winch brake, or by dropping the pawl into the gear wheel on the feed shaft does not afford adequate security against falling; hooks or wedges are needed but blocking devices that prevent the running wheels of the skip from moving will also serve the purpose.

Protection against contact with the stirring mechanism of automatic mixers

Contact with the stirring mechanism in the trough of an automatic mixer should be prevented. As a rule this is done by providing the trough with a cover that cannot be removed, but is so designed that the mixing process can be observed. The cover should also be so interlocked that when it is opened the stirring mechanism automatically stops. This interlocking also applies to the stirring mechanism itself if the arms can be lifted out of the trough while the machine is running. Covers made of grating do not serve their purpose if the spaces between the bars are wide and a hand can reach between them into the trough, for example to try the consistency of the mixture.

(b) ### Installation and operation

Mixing machines should be installed on firm ground so that they are stable and cannot be displaced. If transportable mixers have lugs for taking the load off the wheels they should rest on the lugs. If a pit is dug to accommodate the charging skip its sides should be secured against collapse, and the danger zone along the path of the skip should be fenced. This fencing can be made a part of the machine, a suitable counterweight lifting it when the skip is raised. The pit or the space under the raised skip should not be entered until the skip has been secured against falling.

The operation and maintenance of concrete and mortar mixers should be entrusted only to reliable persons over 18 years of age who are familiar with the machines.

Before any cleaning or maintenance work is done measures should be taken to prevent the machine from being started up again. It is not enough to switch off the electric power; starting up by third persons should be effectively prevented.

2.2 ### Electrical installations

It is wrongly assumed that the really dangerous zone of electricity is clearly marked off by the well-known red sign for lightning. The dangers of electricity are all too often underestimated, and this is especially true of the building industry; not only nearby high tension lines carrying 1,000 volts and more, but also the ordinary tensions of 220-380 volts are dangerous, more particularly because rough handling is an extra cause of electrical accidents.

Voltages over 65 V and currents of 0.05 A should be considered liable to cause fatal accidents because they can induce fatal fibrillation. Other injuries in electrical accidents are burns that can be seen at the points of entry and exit on the body.

It is unquestionable that electrical risks at building sites are factors of great importance; a distinction can be drawn between risks due to working on and near live plant and conductors and the risks of contact voltages.

Precautions for work on or near live plant and conductors

Apart from some exceptional cases, work on or near live parts of electrical equipment should be forbidden. Before work is begun on electrical installations and equipment a responsible person should cut off the current and prevent it from being restored. For this the following measures are necessary:

(1) cut off the current;

(2) prevent restoration of current;

(3) verify absence of current;

(4) earth and short-circuit conductors;

(5) cover or enclose nearby live parts.

This sequence of operations should be adhered to at all times. The current should not be switched on again until the earth and short-circuit conductors have been removed and the workplaces reported safe for the purpose. The report should preferably be in writing. The current should only be switched on again at the order of the responsible person who had it cut off. Agreements that the current should be switched off for a specified time are inadmissible.

When work has to be done in dangerous proximity to live parts the current should be cut off. If for operational reasons this cannot be done the live parts should be fenced off or enclosed. These measures should only be carried out by qualified staff from the power station concerned. Dangerous proximity means more than the space directly within a hand's reach, it varies with the voltage of the line. In addition to direct contact by the body, indirect contact with live parts is possible, and no less dangerous. For instance indirect contact may occur when handling structural steel lengths, placing of reinforcing concrete, handling scaffold parts or long ladders, or transporting tall construction machinery. Precautions will also be necessary in the operation of cranes and excavation machinery near electric lines. The jib must be suitably protected or its range limited. Protection can take the form of wooden stakes. To limit the range of the jib limit switches can be installed that prevent the radius of operation from extending beyond previously fixed bounds. Safe distances should also be provided for when installing the conductors supplying current for the building site. There must be minimum clearances of at least 5 m above the ground, and 1.25 m horizontally and 2.50 m vertically from buildings, scaffolds and the like. On high tension lines there is the risk of a flashover through arcing up to a certain distance even if no contact is made. If heavy equipment is transported alongside or under lines the following precautions should therefore be taken to prevent arcing:

Voltage kV	1-45	Up to 110	Up to 150	up to 220	up to 300	up to 380
Clearance m	1.50	2.00	2.30	2.85	3.10	4.00

In addition the persons employed on the work should be given the most precise instructions on what to do if a machine comes into contact with an overhead line. Anyone in contact with the ground who touches a machine that has become live is in danger, but the excavator or crane operator will not be directly endangered if he remains in his cab. Consequently before he leaves the cab he should break the contact with the line by turning the machine or moving it away.

There is also danger for anyone within the zone of influence of the current, that is near the machine, even if he does not touch it; he should keep still with his feet close together till the machine has left the danger zone or the current has been switched off.

The danger of indirect contact with live conductors exists in circumstances in which they are not directly visible, as for example in work on walls in which conductors are laid under the plaster, when slots or holes are made with metal tools; or in excavating when the excavator catches an underground cable. Electrical conductors should be switched off when buildings are being altered, and before any excavating begins inquiries should be made in the proper quarter whether there are any underground electric cables in the area to be excavated. If there are, the necessary measures should be taken to have the current switched off, subject to the same precautions as have been described for overhead lines.

Measures against the occurrence of contact voltages

A contact voltage occurs when owing to damage to insulation or a defect in an electrical installation, a normally noncurrent-carrying part, for instance the enclosure of the plug and socket connection of an appliance or a machine, makes an electrical connection with the circuit. There is then a path to the body. The voltage between the live enclosure and the earth is the contact voltage. If a person touches the enclosure the current will flow through him to earth. The amount of the current which will pass through a person depends on his body resistance, manner of contact and length of the contact.

On building sites both prerequisites for the occurrence of contact voltages are present, because on the one hand the rough treatment they receive easily leads to mechanical damage in electrical installations and appliances, and on the other hand the worker's position usually provides a fairly good conductor, that is, has a low resistance. This is true both of earthen ground and cement floors, as well as of structural steel. Consequently there are two requirements concerning measures against contact voltages: all parts of electrical equipment for building sites should be built to withstand rough handling and the regulations concerning their installation and use should be strictly observed.

2.2.1 Safety requirements for electrical installations

(a) Distribution switchgear

Every building site must be supplied with electric current from a special supply point. As a rule the supply is arranged through distribution switchgear. The type and size of the switchgear cubicle will depend on the connections to be made. Cubicles should be made of metal which is earthed, or of insulating material; wood should be used only for mounting, lining or framing. It should be possible for the cubicle to be kept closed while it is in operation. Inside the cubicle the wiring plan should always be displayed. By way of additional precautions (2.2.2) the cubicle should be equipped with one or more ground leakage protective devices. If there is only one fault current trip, a fault would cut off the entire current supply, and for this reason at least the power/current supply and the light current supply should each have a separate ground leakage protective device.

(b) Conductors

Stringent requirements should be laid down for the quality and installation of electrical conductors. At building sites strong rubber covered conductors or their equivalents should be used. Exceptions may be allowed for electric tools or lamps, for which medium-strength rubber covered conductors will suffice. However, for both strong and medium-strength conductors the insulation should be treated to increase its resistance to oil and to make it less combustible.

Cables lying on the ground may be damaged by loads dropped on them, by being run over and by crushing, and those at places where they are exposed to exceptionally heavy mechanical stresses should be protected by special measures, such as stringing them overhead. For stringing cables on building sites see also section 2.2. If roads cross building sites overhead lines should have a clearance of at least 5 m at the greatest sag. At the junctions of overhead lines there should also be means of relieving mechanical tension.

If cables cannot be strung overhead at road crossings, cable covers can be used to provide the necessary mechanical protection for them. Since rubber-covered conductors are not cables they should not be buried in the ground. Where necessary these rubber-covered conductors should be protected by wooden casing, conduits or the like.

Frequent causes of electrical accidents are faulty joints and connections at plug and socket appliances, switches, motors, and switchboards. Such joints and connections should be made only by electricians. At the connection points they should be relieved of tension and compression. At the points of entry to machine housings the leads should be protected from damage and kinking by rounding the edges or providing bushings. Since current can leak to the housing as the result of unravelling of a single wire, the ends of conductors should be soldered or fitted with cable shoes. Damaged or torn conductors should not be patched up with insulating tape and should not be used again until they have been properly repaired.

(c) <u>Plug and socket connections</u>

The variety of types and forms of plug and socket connections leads to improvisations that always entail a risk of accident. Consequently only standard connections should be used on building sites. Different plug and socket configurations should be used for different voltage or current ratings. The connections should have insulated and watertight covers. Only one flexible cable should be connected to any plug. Plugs and sockets should be so arranged in the lighting circuit that when they are disconnected the plug can never become live. They should not be disconnected by pulling on the cable for this would nullify the tension-relieving arrangement. If machines have to be repaired it is not enough to switch them off; they should be disconnected from the mains by pulling out the plug. This is particularly necessary when a loose wire can make contact with the housing.

(d) <u>Lamps</u>

Lamps at building sites comprise hanging lamps, standard lamps and hand lamps. Lamps should only be put to the use for which they were designed; for instance a hanging lamp should never be used as a hand lamp. Lamps on building operations should be of types that can stand up to rough handling and should also be weatherproof. Strict requirements are particularly necessary for handlamps: they should be both effectively insulated and waterproof.

(e) <u>Electric tools</u>

Electric tools such as drilling machines, grinding machines, portable woodworking machines and concrete vibrators should only be used in installations with pressures up to 380 V and with the usual protection against contact voltages. Every electric tool should have a switch for switching it on and off. Portable machines for wet grinding should only be operated on a low voltage or with double insulation. This also applies to all electric tools that are to be used in highly conductive interiors such as boilers and piping, since here the risk of contact voltage is particularly serious.

(f) <u>Electric machines</u>

Electrically driven machines should be so constructed that all electrical parts are protected against contact with tools and the penetration of small foreign bodies and water. Current should be supplied from a special source, usually a switchboard. This applies, for instance, to small concrete mixers, which should not be directly connected to wall outlets of fixed installations for example in domestic property.

Switches for starting and stopping building machines should cut off the current on all poles. The switch should be in an easily accessible place and within easy reach of the operator. It should be ensured, either by the design or by suitable protection, that live parts are safe from any accidental contact.

2.2.2 Additional precautions against excess contact voltages

Owing to the stresses imposed on electrical installations by the special conditions and heavy work on building sites it is not to be expected that technical requirements concerning the design, installation and use of electrical equipment will be effective indefinitely and so prevent contact voltage from ever occurring, hence, additional protective measures are necessary. They include double insulation, low voltage, protective isolation and ground leakage protective devices.

(a) Double insulation

With double insulation all accessible current-carrying parts of an electrical installation or a piece of equipment are securely and permanently wrapped in insulating material, or by means of internal insulation are isolated from all parts that in the event of a fault could directly become live. For constructional reasons, the use of double insulation as an electrical precaution is almost confined to the conductors of electric tools, transformers, small appliances and building-site switchboards. When assessing the value of double insulation as a safeguard it must be remembered that care has to be taken to prevent damage to the insulation, and that if damage does occur the equipment concerned should not be used again until it has been properly repaired.

(b) Low voltage

When low voltage is used as a protection, electrical appliances are operated at a rated voltage not exceeding 42 V, and a protective transformer or a converter is inserted in the circuit. Low voltage is used mainly for electric tools and handlamps. It should not be possible to insert plugs for low voltage into sockets of installations for higher voltages.

(c) Protective isolation

With protective isolation an isolating transformer or a motor generator isolates the consuming appliance from the supply system. However, only consuming appliances taking not more than 16 A current should be connected to an isolating transformer. Since protective isolation is ineffective when a fault to earth occurs on the appliance side, special care must be taken to avoid damage to conductors and plugs. All things considered, the value of protective isolation in building operations is doubtful.

(d) Ground leakage protection devices

This type of protection has proved particularly useful at building sites, largely because ground leakage protective devices have been developed that meet the special requirements of these sites. When a fault current occurs that reaches the tripping strength of the device, it acts within 0.2 seconds, cutting off the current from the appliance, machine, etc. on all poles. The ground leakage protective device, which is based on Ohm's law, takes care of an incipient fault. As soon as it is so large that the resistance to earth of the appliance results in a voltage drop of 65 V, a magnetic switch opens and makes the appliance dead.

Only one ground leakage protective device is needed to protect an entire installation, provided that the tripping capacity of the device is adequate for the load. However, individual circuits or consuming appliances can be protected by separate devices. As a general rule, a number of circuits each protected by its own ground leakage protective device is preferable in an arrangement where the whole installation is not put out of service as soon as one appliance develops a contact voltage. The efficacy of the device depends on the provision of perfect insulation and earthing. For the earthing of the device at building sites, lengths of angle iron are most frequently used: for these a minimum cross-section and protection against corrosion are prescribed.

Before operations begin the earthing should be tested by a competent person. If ground leakage protective devices are used, separate earthing of machines and earthing of rail tracks of tower cranes, is always desirable and improves the general earthing. However, the earthing device must have the right value, and consequently when it is being tested all other earth conductors should be disconnected.

2.2.3 Testing of electrical installations

The safety of an electrical installation depends to a decisive extent on its condition. It is not enough to have equipment properly installed by an electrician. The different items should be carefully handled and properly used and the installations should be examined regularly to ensure that they are in perfect condition.

The examination should include a thorough inspection of all parts of importance for protection purposes, and the measurements required to show whether the protection is effective. During the inspection particular care should be taken to see whether the protective conductor has the prescribed cross-section, is properly laid without discontinuity, is carefully connected, and over its whole length is identified in the prescribed manner. Examination of conductors will be mainly concerned with the condition of the insulation, protection against mechanical damage, and the perfect condition of connections and points of entry.

Monthly tests of the efficacy of the ground leakage protective devices usually installed on building sites are indispensable.

2.3 Lifting appliances, building hoists, lifting gear

The importance of lifting loads in building operations is evidenced by the variety of lifting appliances and hoists used in these operations. The main difference between these two types of equipment lies in the way the load travels; on lifting appliances it hangs freely while in hoists it moves between vertical or nearly vertical guides, sometimes with the assistance of a winch.

The types of risk are as numerous as the types of lifting appliances and hoists: they may come from constructional defects or from faulty operation. Consequently, exceptionally strict safety requirements should be imposed concerning the construction, equipment and operation of, etc., lifting equipment.

Lifting appliances

Lifting appliances include the various types of crane, electric hoists, and of course also the simplest appliance, the winch, which is used with pulleys.

2.3.1 Winches

The most important component of any lifting appliance or hoist is the winch. The technical safety requirements concerning winch design are discussed generally in the following sections, and if a few operations entail special requirements these are explained in the discussion of the type of lifting appliance or hoist concerned. Irrespective of their type, all winches consist of a barrel or drum, a rope, winding mechanism and a brake. According to the type of drive a distinction can be drawn between hand-operated and power-driven winches, and the latter can be divided into winches with a one-way or continuous drive and winches with a reversing drive or coupled winches and reversing winches (2.3.1(f)).

(a) **Identification plate**

For identification every winch should have a plate giving the maker's name, type of construction and maximum permissible load. With power-driven winches the proper cable diameter and the breaking strength of the wires in the cable should also be indicated so that the right cable can be chosen if a cable has to be changed. Because of the generally imposed restrictions on the raising and lowering speeds of cranes and building hoists, plates on power-driven winches should show the maximum permissible operating speed, and hence the maximum number of revolutions of the driving shaft.

(b) **Drums**

As will be explained later (2.3.1(d)) wire ropes as load-bearing elements of winches are exposed to varying stresses. One of these is caused by the bending of the rope as it winds on the drum or passes over the pulley. To limit this bending stress the diameters of the drum and the pulley should be suitable for the diameter of the rope, and in fact the diameter of the drum should be at least 20 times, and the diameter of the pulley at least 22 times, that of the rope.

Another stress, and with it damage to the wire rope, occurs when the rope winds irregularly on the drum and the turns are laid diagonally over each other, causing crushing. On power-driven winches at which only one layer of rope is wound so that the drum must have a certain length, grooves on the drum prevent irregular winding of the rope. The radius of the grooves should be suited to the diameter of the rope. If the rope is wound in a number of layers on the drum, by suitably placing the winch or making other arrangements it can be ensured that the rope is wound evenly in parallel turns and is hence protected against damage.

As the rope might be damaged by sliding off the side of the drum, which would affect the smooth raising and lowering of the load, the drum is provided with rims at the edges; these should be high enough to be effective when the rope is fully wound. This means that the top layer of rope should be at least two rope diameters below the top of the rims. Consequently unlimited lengths of rope cannot be wound on drums, even if, for the lifting height, a rope is needed that is longer than the drum can safely take: in such cases a winch should be provided that can take the necessary length of rope.

Drum design should also ensure that the rope can be securely fastened. For this purpose the drum should have a special device made so that the fastening cannot be weakened. For means of attachment a wedge, strap or clamp device may be used. The fastening should be so arranged that it is visible when the rope is unwound so that the winch operator can always notice any defect in it. An extra hold on the rope is ensured by the requirement that at least two turns shall always be left on the drum after unwinding. On electric winches compliance with this requirement can be ensured by providing a limit switch.

It is also important that the rope should be fastened on the drum in the direction of rotation for lifting. If it is fastened in the opposite direction it might kink and be damaged. For this reason the winding direction of the rope on the drum should be indicated by an arrow.

(c)   Wire ropes

Requirements concerning the specifications of wire ropes for lifting appliances and building hoists are not the same in all countries so that here only generally applicable recommendations will be made. The types and sizes of ropes for the different kinds of lifting appliance are governed by national regulations or standards and the particulars given on the winch plate by the manufactuer. This applies both to the first rope for a new winch and to replacements when the time has come to withdraw a used rope. The working life of a winch rope depends on the static stress, the bending stress as the rope runs over the drum and the pulleys, and the type of rope, a matter that has to be taken into account when choosing a rope. Ropes wear both internally and externally. As it runs over the drum and pulleys the wires bend and slide against each other, causing fatigue in the rope and wear on the wires inside. There is also the external wear and tear on the rope which with the internal damage may lead to breakage of individual wires. It is on these considerations that requirements applying to ropes are based. The requirement concerning the mutual adaptation of rope and drum has already been discussed. For winches, use should only be made of wire ropes that have been galvanised or given equivalent protection against rust, and possess a safety factor of six; that is, the breaking strength of the rope must be six times the maximum permissible load shown on the winch plate.

Wire ropes for winches should be composed of at least 114 wires and not more than one core of hemp or similar material. Wire ropes should be in one length; splices, which should be properly made, should only be used at rope ends for fastening, and should not be allowed elsewhere. For making eye splices, thimbles should be used with a radius of curvature at least three times the rope diameter. If jaw clamps are used, the bearing part of the rope should be fastened to the loose end by at least three clamps. The loops of the clamps should press against the loose end. The distances between the clamps should be at least equal to five times the diameter of the rope.

Wire ropes for winches need constant care and attention. The moment for discarding them will depend on the number of broken wires. The table below may serve as a guide for deciding when to discard a rope.

DIN 655 DIN 656 Number of wires in rope	Wire ropes according to DIN 6895 Number of		Number of visible broken wires at time of discard			
	Wires in rope	Wires in outside strands	Crosslay over a length of		Ordinary lay over a length of	
			6d	30d	6d	30d
	10 x 10 = 100	6 x 10 = 60	5	10	2	4
	18 x 7 = 126	12 x 7 = 84				
6 x 9 = 114	10 x 12 + 36 = 156	6 x 12 = 72	8	16	3	6
	36 x 7 = 252	18 x 7 = 126	15	30	5	10
8 x 19 = 152			18	36	6	12
	4 x 7					
	+5 x 20					
	+7 x 24 = 296	7 x 24 = 168	25	50	8	16
6 x 37 = 222	6 x 37 = 222		30	60	10	20
8 x 37 = 296			40	80	12	24

Wire ropes should also be taken out of use as soon as a strand breaks, or bulges, kinks or other serious damage is noticed. Heavy rusting and badly worn wires are also reasons for discarding ropes.

(d) Rope pulleys

The diameter of all rope pulleys used with lifting appliances should be a certain multiple of the rope diameter, namely 22 times this diameter.

With certain types of lifting appliance the dimensions of pulleys will depend to some extent on the effects of bending and reverse bending of the rope, which may necessitate a larger pulley diameter. This point will be dealt with in the sections on tower slewing cranes and building hoists.

All rope pulleys should be so made that both legs of the rope running over them are in the same plane. Consequently the idler pulley over which the rope from the winch is led upwards should have a swivelling mounting, so that it can keep aligned with the rope at any of its positions on the drum. The same result can be attained if the idler pulley can move on a horizontal axis. To prevent the rope from jumping out of the pulley, the pulley should have a hoop that will also stop the rope from whipping out if the pulley breaks.

(e) **Winding mechanism and brakes**

All winches used to raise loads should be so made that the load cannot descend inadvertently. The devices for this purpose should work automatically and smoothly, be protected against unauthorised interference and not be put out of action by bad weather. All winches should also have a reliable brake adequate for the lifting capacity. These two basic requirements - for security against accidental descent of the load and for a brake - are the main factors determining the design of the different types of winch, namely, hand-operated, power-driven with a one-way (continuous) drive and power-driven with a reversing drive, as will be described in section 2.3.1(f). The controls of winches should be easily and safely accessible. Near each control should be a notice clearly indicating their purposes and method of operation. The designs should prevent any inadvertent movement or displacement of the control lever.

A general requirement for brakes is that they should be made of suitable material and work faultlessly in normal working conditions even if they are used intensively. They should act without jolting or vibration and without much delay. All parts that might become loose or detached should be properly secured. Brakes should be adjustable so that any play resulting from wear and tear and capable of impairing smooth working can be eliminated by simple means. Brake bands which can be adversely affected by external influences such as water or oil should be protected against these influences. Bands should be designed for a braking moment at least 1.25 times the load moment.

(f) **Types of winch**

Hand-operated drum winches

Hand-operated winches where the drum is rotated either directly or through a toothed transmission mechanism and gearing by a crank handle have two main sources of danger that should be eliminated in the design. First measures must be taken to prevent the engaged transmission from springing out as a result of inadvertent displacement of the gear shaft so that the load descends freely; and second the handle should be prevented from kicking back, a cause of very serious accidents. This means that the control lever for engaging and disengaging the transmission should be effectively protected against displacement. Pawls alone are not sufficient protection because winch operators often forget to use them.

To ensure safe raising and lowering of the load with every type of transmission between the handle and the winch shaft and so prevent dangerous kick backs, hand-operated winches should be equipped with a built-in kick-back preventer. The simplest protection against kick-backs consists of a ratchet wheel in which a pawl is inserted while the load is raised. However this protection can only be used for raising because it must be removed for lowering. Consequently kick-back preventers should be used that allow the pawl to remain engaged during lowering. For this, between the handle and the ratchet wheel, or between the driving shaft and the ratchet wheel, there should be a clutch that works automatically in such a way that while the load is being raised or held the pawl is engaged, but the handle is disengaged while the load is being lowered. In this way the handle is kept still during lowering and the load is held by the brake.

There are other devices for preventing kick-backs, for instance load-pressure brakes and wormsheel drives.

On winches with two handles one handle should not disengage the kick-back preventer of the other. As further protection against the load falling by its own weight, hand-operated winches should also be equipped with a brake suited to the lifting capacity; as a rule this will be a band brake. The band lies round the brake drum and a weighted or spring-loaded lever presses it against the drum with the necessary friction. If a load is to be lowered the brake is released by a manually operated lever. It is important that when the control handle is released the brake is automatically reapplied and holds the load.

Special attention should be paid to the form of the handle grip on hand-operated winches. The revolving sheath placed over the grip to prevent the surface from being overheated by friction during rotation should be so formed that there is no nip point that could cause hand injuries.

Power-driven winches with reversing drive

These winches are driven by an electric motor which acts through the drive on the winch drum so that the upward and downward movements are controlled. The changeover between raising and lowering is thus effected by reversing the direction of rotation of the motor. Winches of this type are used on all tower slewing cranes, on some kinds of building hoist, on charging mechanisms, and on low-power electric hoists, and also as electrical erection winches. Like all other winches, power-driven winches with reversing drive should be equipped with a reliable brake suited to the lifting capacity, and such that the operator cannot produce any additional braking effect. Any addition to the braking effect by the operator would overstrain the whole winch including the lifting gear and might cause serious injuries and damage. Most brakes are double shoe brakes which force the shoes against the brake drum by springs thus immobilising the winch drum and the load. The brake is released by electrically controlled magnetic fields that remove the shoes from the drum by opposing the force of the springs. As soon as the current is cut off from the motor, or fails because of a fuse blowing, supply breakdown or other incident, the brake release acts and the brake is applied automatically. Special kick-back protection is not required for power-driven winches with reversing drive because with magnetic braking kicking back is prevented by the nature of the brake. Similarly, on these winches speed governors limiting the lowering speed can be dispensed with since this function is performed by the motor. It is merely required that the maximum permissible speed of the motor's rotation should not be greatly exceeded.

Power-driven drum winches with one-way drive

These winches are driven by an electric motor or an internal combustion engine that always rotates in the same direction. To raise the load the motor is connected to the winch drum by a clutch. To lower the load the clutch is withdrawn and at the same time the brake holding the drum is released by means of a manually operated lever.

This type of winch is widely used in building operations. Usually the main brake is a band brake. Here again the band lies round the brake drum and is forced against it by a weighted or spring-loaded lever with the necessary friction. For lowering, the brake must be released by a hand lever. Requirements for this arrangement are, first, that when the brake lever is released to brake should automatically be applied, and second, that it should not be possible for the operator to supply additional braking force that exceeds the permissible maximum.

On winches with a one-way drive the runaway preventer performs an important function. It has to prevent the load from descending accidentally especially at the moment when the winch is being started up or stopped, that is when the brake is off. The runaway preventer may be either a differential brake or a catch. With a differential brake both band ends are so fastened to the brake lever that the compressive force of the band is always opposed by arms connected to the pivot of the brake lever. With this arrangement, when rotating in the raising direction the braking movement is reduced almost to zero without the brake being released, while in the opposite direction the brake is automatically applied without the application of any external force. Thus the winch drum can only run back when the brake is released by means of the lever.

Catches, of which there are several types, act like free wheels. As soon as the raised load is brought to a halt by the disengagement of the clutch the catch is forced against the inside of the brake ring, on which the brake will remain blocked. The action is the same if the winch drum runs back accidentally when for any reason the electricity is cut off, for instance because of a supply failure, a clutch failure, or tearing of a belt. If the load is to be lowered the brake is released. Then the brake ring with the winch drum also blocked by the catch moves freely and the load can be lowered almost at the speed of free fall.

Power-driven winches with one-way drive may be built with one or two operating levers. With the two-lever type one lever is for the clutch and the other for the brake. Direct switching from lowering to raising cannot be made impossible by constructional measures. It can only be prevented by correct operation, for the operator must first put the brake lever at zero before engaging the clutch with the second lever. With the one-lever type the same lever serves both to engage and disengage the clutch and to release the brake. Fundamentally, the operating arrangement is that the zero position, that is the braking position, is in the middle. Arrangements with the lowering position in the middle are inadmissible. There are various guide systems for operating levers. With one-way switching, to prevent the lever from accidentally moving from one position to another, an automatic stop is provided at the braking position, and it has to be taken off before the lever can move again.

Without an additional device, on power-driven winches with one-way drive, the load could descend at the free fall speed when the brake is released. Since at such a speed quickly stopping the load by applying the main brake would cause severe jolting that might damage not only the winch but the whole lifting appliance or hoist with its lifting gear, automatic speed governors are required for power-driven winches, at least if the lifting capacity is above a certain minimum, which may be taken as about 600 kg. These speed governors are usually based on the centrifugal principle; when a certain speed is reached weights are forced against the edge of the drum and so reduce its speed to 90 m/min. Speed governors are required both for cranes and for hoists if they operate with winches with a one-way drive.

(g) Installation of winches; operator's stand

When winches are built-in components of lifting appliances or hoists, including cranes and charging appliances of mixers, the manner of their installation is determined by the construction of the lifting appliance or hoist. In all other cases winches should be so installed that they are adequately secured against displacement and lifting off their base whether by vibration or by the pull of the load. Consequently provision of a special anchorage or other precautions may be necessary. The winch should be perpendicular to the rope as it runs on and off the drum, and the angle to the horizontal made by the rope between the drum and the bottom pulley should not be large. Care should also be taken that the rope can wind smoothly on the drum, and for this the distance of the idler pulley from the drum should be at least ten times the length of the drum. Over this distance the rope should be fenced or enclosed so that no one can be caught by it or stumble over it.

The drive and the operator's stand of winches should be roofed over at a height of about 2 m as protection against falling objects and the weather. In no case should the operator's stand be in the danger zone of the lifting gear or the load. From his stand the operator should have an adequate view of the loading and unloading points so that at any moment he can see where the load is. Where, for local reasons, such a view cannot be provided, measures must be taken for the arrival of the load at unloading points to be indicated, for example, by distance marks on the rope. It is better, however, to have a limit switch that automatically stops the winch when the load reaches the unloading point. Limit switches may act either mechanically through a rope to the operating lever, or electrically by cutting off the current, and they are valuable means of preventing over-running. For communication between the operator's stand and loading and unloading points a signalling installation such as a telephone or bells should also be provided.

(h) Operation of the winch

Persons operating winches should be suitable for the job, reliable, not under 18 years of age, and properly trained. According to the nature of the lifting appliance additional qualifications may be required, more especially concerning training (crane and hoist operators). Before starting the day's work winch drivers should satisfy themselves that the winch is in safe working order. If any defects are noticed they should be reported to the management or to a responsible employee.

Winches should not be loaded beyond the maximum shown on the plate. Persons should not be carried on the load or the lifting gear. The operator should not leave the winch while a load is suspended.

At least once a year all parts of the winch should be tested by a competent person. Additional requirements are applicable to cranes and hoists.

2.3.2 Simple blocks and tackle

The simplest lifting appliance at building operations is the block and tackle in which the lifting rope runs over a pulley block fastened to an overhead beam or gallows (Fig. 9). The load is raised by a hand-operated or power-driven winch (1.3.1). With this type of lifting appliance the dimensions and fastening of the beam are all important. Near the point of attachment of the pulley the force acting on the beam is much more than twice the load on the rope. For this reason the beam should be chosen for the maximum static load multiplied by a shock factor (S). For hand-operated winches S should be 3.0, for power-driven winches with electrically controlled lowering S should be 5.0 and for power-driven winches with mechanically controlled lowering, 8.0.

The method of fastening the beam will depend on structural conditions on the site. It should be prevented from tipping over its front support. If it cannot be anchored to a fixed part of the building with steel hoops a counterweight should be fastened to the back end. Allowing for the shock factor the counterweight should be calculated for a safety factor of at least two against tipping, and should be securely fastened to the beam. It is thus not enough to lay loose stones, rolls of cardboard or the like on the end of the beam. If ballast is used as a counterweight a large enough bin should be provided. The pulley should be secured against displacement or unhooking by being clamped to the beam.

The roofer's sheerlegs are a special type of block and tackle. The front end of the beam is laid on a yoke so that raised loads can be taken off directly under the suspension point of the pulley. The yoke should be firm, and to ensure this it should be securely fastened to the beam, and its feet should be anchored to the roof or clamped to the back end of the beam in a triangular arrangement. At the top of the sheerlegs the beams should be secured against

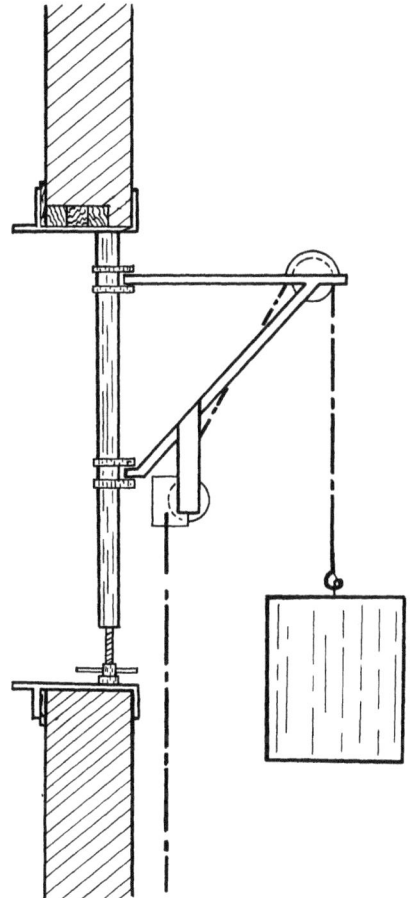

Fig. 9
Window mounted block and tackle

displacement. At the front where a person might fall railings 1 m high should be fitted. With low gallows the railings may have a removable section in the middle to allow the suspended load to be pulled in. To prevent the load from swinging it is useful to provide a tag line.

2.3.3 Pivoting gallows

This type of lifting appliance consists of a mast placed on the front of a building or a scaffold and carrying a pivoting beam at the top. The lifting rope runs over the pulley on the beam to the foot of the mast, from whence it is taken over an idler pulley to the winch.

The mast should be calculated for the maximum load to be moved. The dangerous part of the cross-section is the top. The mast should stand firmly on a load-distributing base and be slightly inclined towards the building. The foot should be firmly secured to the base so that it cannot move. To prevent buckling the mast should be anchored to the building at every storey in a manner that avoids tensile and compressive stresses. Anchorages should not be more than 3 m apart. At the top the mast should be tied to the building with wire ropes. The beam should be so fastened and secured to the mast that it does not fall down. The collars of the beam should be laid tightly round the mast and screwed on (with tensioning chains or screw bolts with double nuts), and secured against falling down. If any packing is needed it should be inserted in the collars from above. The mast should be so arranged that the lower idler pulley is at least 2.50 m above the scaffold floor.

If pivoting gallows are installed on a scaffold, the scaffold should be built to withstand the resulting stresses both vertical and horizontal. With a maximum lifting capacity of the hoist at 600 kg, the load on the scaffold should be taken to be 300 kg/m².

In the use of a pivoting gallows installed in front of a scaffold there is also the risk of the load catching in the scaffold. To avoid this, on the outside of the scaffold a screen of boards should be built preferably reaching to 1 m above the scaffold floor, so that the top of the screen also serves as protection against falls of persons. The load should be lifted above the top of the screen.

Among varieties of the lifting appliance with a pivoting beam are the so-called gallows hoist and the window crane.

With the gallows hoist the pivoting outrigger is fitted on a 3 m high wooden pole that is installed on a storey ceiling or a scaffold floor to which the load is to be raised. The mast is fastened to a base plate that should be securely anchored to the ceiling or floor.

With the window crane the steel crane post is fixed in the window opening. To fix the crane in position strong blocks are placed against the lintel and the sill and they hold the U-shaped head and foot pieces of the crane post. With this arrangement, if the idler pulley is not at least 2.50 m above the floor, it should be screened to prevent a hand from being accidentally caught in it.

2.3.4 Low-power electric hoists

Unlike other lifting appliances these electric hoists have a winding mechanism consisting of a double electric winch at the top loading point that acts as a crab running on the bottom flange of a steel girder serving as the cross-bar of a steel trestle.

Whenever possible this trestle should be erected on massive flooring. Having regard to the weight of the loads to be lifted, and especially with floors that have just been concreted, the legs of the trestle should rest on load-distributing bases. If an electric hoist is installed on a scaffold, the scaffold should be strong enough to bear the maximum permissible load of the hoist plus the dynamic stresses that will occur. Since the weight of the unloaded crab alone is enough to tip the trestle over, especially if the crab track extends beyond the trestle and so overloads one side, it is essential to anchor the trestle securely to its supporting surface. If because of conditions on the spot this anchoring cannot be provided, a sufficiently heavy counterweight must be used. Its weight should be given by the hoist manufacturer in the operating rules, and in no case should a smaller weight be used. If loose ballast is used sufficiently large containers should be provided.

At these electric hoists protection against falls, usually by railings, should be provided at the top landing. The hoist operator should never remain unprotected at the front edge of the landing.

Measures should also be taken to prevent the moving load from catching in the building or a scaffold, for instance by using a tag line.

Cranes

2.3.5 Rotating tower cranes

Among the lifting appliances used on building sites, the rotating tower crane is distinguished by its ability to perform three kinds of transport operation: it picks up material at the store place or preparation point and moves it horizontally and vertically direct to the workplace or installation point and then deposits it there without any other form of transport being required - provided that the crane is properly used.

Accident frequency in handling loads is very high, and any simplification of transport operations is a contribution to safety. It may therefore be said that the rotating tower crane is a notable contribution to accident prevention but this is true only of some aspects of crane operation for it also presents many risks. It is therefore important to know these risks and combat them.

Crane accidents may be due to widely differing causes which not infrequently operate in conjunction. They may include defective stability of the crane caused by faulty calculation or faulty construction, and in the case of track-mounted cranes, defective track. Other accident causes include faulty operation, improper movement of loads, and poor slinging. Crane operation entails a multiplicity of different precautions if it is to be safe.

(a) <u>Crane construction generalities</u>

Rotating tower cranes have a latticework tower or a mast with a jib at or near the top. Different types of tower crane are distinguished by the form of the jib, as follows:

<u>Needle jib cranes</u> have a jib on the tower moved by a rope that runs over the horizontal or inclined top of the tower.

<u>Articulated girder jib cranes</u> have a girder jib at the top of the tower, adjustable on a small arm.

<u>Cranes with crabs</u> have a horizontal jib on which a crab runs. The lifting rope is suspended from the crab so that a raised load can be transported horizontally by moving the crab.

<u>Bell jib cranes</u> on which the top of the tower pivots as well as the jib and a second jib carrying out the counterweight. The lifting rope runs over the top of the tower.

(b) <u>Requirements concerning soundness and stability</u>

All parts of the lifting gear and mechanical and electrical equipment of tower cranes should satisfy generally recognised technological requirements, as laid down in the rules of the Fédération Européenne de la Manutention for the calculation of lifting appliances.

Crane calculations should include unambiguous particulars of the main dimensions (shown in a general scale drawing), the weight of the most important parts, the nature of the material used, the assumed stresses, the types of stress, the assumed maximum stresses in structural parts and connections, the safety factor in critical loading conditions with particular reference to kinking, buckling, fatigue, etc., and the extent of sagging, in so far as these matters have any significance for the construction or operation of the crane and for its stability.

The stability of cranes used on building sites should be ensured in all normal working conditions and also in stormy weather, whether they are loaded or not.

(c) <u>Construction and equipment</u>

<u>Crane plates, crane types, counter-weights and other notices</u>

On tower cranes the particulars indicating the type of crane should be shown on a plate, as follows:

manufacturer's name;

crane type;

factory number;

year of construction;

in the case of tower cranes subject to type tests, the test mark.

The maximum permissible load should be marked conspicuously on the crane.

On cranes with a movable jib, the maximum load should be shown for the different radii, and at least for the greatest, least and middle radii. Tower cranes on which the jib radius can be changed under load must also have an indicator showing conspicuously the radius at any moment and the corresponding maximum permissible load. The plate should be so placed that no only the crane operator but also the slinger and the foreman can always see the radius and the corresponding maximum load. If a tower crane has more than one operator's stand the indicator should be in the operator's field of vision at every stand.

On jib cranes easily legible maximum load tables should be affixed.

Since the stability of a crane depends partly on the weight of ballast allowed for in the stability calculations, the weight and the position of the ballast should be given in the operating instructions. If the height of the tower and the type of jib may make more than one counterweight or load of ballast necessary these requirements should be shown in tables affixed to the crane.

Lastly, the instructions for operating the crane should be displayed in an easily legible notice. The notice should be weatherproof if it is in the open air and exposed to the weather, as when tower cranes have no operator's cab but are operated from a stand on the ground. However, the display of operating instructions in the operator's cab or at his stand is no substitute for the training he needs. The chief purpose of the notice is to enable the operator to look up the instructions at any time, especially if he is in doubt.

Winding mechanism, wire ropes

Generally speaking the winding mechanism of tower cranes for raising, rotating and moving the jib should satisfy the requirements for winding mechanism set out in section 2.3.1(f).

The power-driven mechanism for raising loads or adjusting the jib should be so designed that lowering the load on the lifting gear or the jib is only possible when the power is in use when the load or the lifting gear is held by the winding mechanism for the whole duration of lowering. This requirement concerning the construction of winding mechanisms excludes the use of clutch winches on tower cranes. If the load is lowered with the current off the rated speed of revolution of the motor should not be exceeded.

Wire ropes used for lifting should comply with the requirements set out in section 2.3.1(d). They should not be loaded beyond 1/5 of the actual breaking strength given by the manufacturer. Wire ropes not used for lifting and not running over pulleys, such as those used for guying the crane should not be loaded beyond 1/4 of this strength.

Special attention should be paid to the fastening of the lifting rope to the hook. At this point the rope is subject to heavy and changing stresses and bends especially while the load is swinging or being set down. The resulting crushing will gradually destroy the inside of the rope so that it may break before the normal time for discarding it.

Electrical equipment

The requirements described in section 2.2 on electrical installations and equipment apply to the electrical equipment of tower cranes. In addition there should be a crane switch by which movement of the crane can be stopped from any operator's stand by cutting off the current on all poles. With remote control, it should be possible to cut off the current immediately from the control point.

Flexible electrical cables should be led over a drum that automatically winds and unwinds them during travel. For this purpose spring-loaded drums or power-driven drums can be used. At the drum the cable should be protected against kinking by guiding devices. A cable should never be led over sharp edges.

Protective measures for the electrical installation of the crane should extend to the rail track: for instance the rail lengths should be properly electrically bonded and earthed.

Operator's stand or cab and means of access

Any mistake in the operation of a crane may impair its stability and care must be taken that the operator's stand is so built and equipped that he can operate it safely and easily. The stand should be so arranged that whatever the position of the crane the operator's view of the area of operations and the load is never restricted. Unless the stand is on the ground or near it a cab should be provided. Cabs should:

by sufficiently spacious;

protect the operator against the weather (rain, heat or cold);

be heated in cold weather;

be well ventilated;

be well and suitably lighted;

be safely and easily accessible;

be provided with a seat;

having windows that can be cleaned safely.

A good field of vision can be ensured both by suitable window design and arrangement, and by suitable arrangement of the crane controls. The operator should not have to lean out of the window to watch the load as it moves.

Starting resistances and rheostats should not be used for heating cabs. Open heating coils are inadmissible because they may cause fires and burns. A suitable plug and socket connection should be used to supply heating appliances.

The parts of the drive and the electrical equipment installed in the cab should be so arranged and protected that the operator cannot inadvertently touch them. The cab should be built of fire-resistant material and be equipped with a fire extinguisher that is easily accessible. The addition of a cab however makeshift on a crane tower cannot be allowed since it would change the static factors governing the stability of the crane.

On tower cranes it must be possible to reach the operator's stand safely by way of platforms, gangways, stairs, ladders, etc. Persons should not climb on to the crane over the tower latticework. Ladders should afford a foothold at least 40 cm wide and 50 cm deep. The rungs should be at least 16 cm apart. Vertical or nearly vertical ladders over 6 m long should be equipped with hoops as back rests starting 2.50 m from the bottom and spaced not more than 0.9 m apart. They should leave a clear space not more than 0.7 m in width. If the hoops are joined together by vertical bars they may be up to 2 m apart. Hoops may be dispensed with when equivalent safety is ensured by other means.

Ladders over 10 m long should be provided with intermediate platforms. Ladders leading to a platform or a gangway should have at least one upright extending 1 m above the top landing unless safe stepping on and off is ensured by other means. If persons can fall off working platforms they should be provided with standard railings. The floors of gangways, working platforms, etc., should not be slippery, and should be so made that water can easily flow off.

Safe means of access should be provided wherever they are needed for erection, maintenance and other work. If the pulleys at the tip of the jib cannot be safely lubricated from the ground by lowering the jib or other means, the jib should be provided with two-rail fencing and a toeboard. It is preferable however to lubricate these pulleys from the top of the building by lowering the jib. A gangway on the jib is not needed if the pulleys have bearings that only need to be lubricated at fairly long intervals.

Counterweight jibs on which ballast is placed after they have been erected should have a gangway at least 50 cm wide and equipped on both sides with two-rail fencing and a toeboard.

Other precautions

Safety in crane operation is also furthered by the following precautions:

proper construction of track-wheel rims;

lugs to support the crane if a wheel breaks;

protection against wind pressure;

brakes for the travelling and rotating motions of track-mounted cranes;

crab safety devices for jib cranes with crabs;

overload preventers and various limit switches.

The wheels of track-mounted cranes must have rims at least 15 cm high. Since rims can easily be worn down by a bad track the wheels should be changed when necessary. The lugs should prevent the crane from overturning if an axle or a wheel breaks or a derailment occurs. Every track wheel should have a lug not more than 20 mm above the top of the rail.

Cranes that run on rails should be protected against inadvertent movements whether due to wind or to travel. Protection against displacement by wind can be afforded by rail clamps, spindle brakes that act on the rails, wheel locks or brake shoes. Brake shoes should be fastened to the crane and secured against accidentally falling. Wind protection is required for both directions of travel and should be diagonal.

If the travelling mechanism is power-operated, the brakes for the travelling motion should be automatic and work smoothly. Sudden hard braking can cause the crane to rock or to overturn. Hence it is not permissible to change the direction of travel abruptly, that is, to stop it by reversing the motor. This applies equally to the rotating motion, and consequently in addition to a brake for the travelling motion there should be a smooth-working brake for the rotating motion, but it should not be applied while the crane is idle. In fact the crane should then be able to move into the direction of the wind like a weather vane. If this were not possible there would be a risk that the crane would be overturned by the wind. If, however, there is a possibility that the jib will be blown against a building or a scaffold, it should be lowered or anchored to a fixed part of a building in such a way that it can withstand tensile and compressive forces.

One of the most important appliances for preserving the stability of a tower crane is the overload preventer, but it does not do away with the need for other measures for this purpose. Its function is only to prevent the crane from being overloaded. When a slung load is heavier than the maximum permissible load for the crane the preventer stops the crane by switching off the lifting mechanism and the mechanism for adjusting the jib. There are different models of this safety device - mechanical, electrical, electro-mechanical and electro-pneumatic. An overload preventer should act if there is an overload of 10 per cent or more. When it has acted, the jib, except on tower cranes for horizontal transport, should be blocked, but it should be possible to lower the load so that it can be set down. With jib cranes equipped with crabs, it should be possible to bring back the crab.

Limit switches should limit:

the highest and lowest positions of the hook;

the terminal positions of movable jibs;

the terminal positions of crabs on jibs.

It is not necessary to have a limit switch for the lower terminal position of the jib if it cannot be moved under load. After the switch has acted it should be possible to make movements in the opposite direction.

Tracks, installation, maintenance, clearances

A well-built track is a good foundation for a crane and thus an important factor in safe operation. Tracks should be so laid that a tower crane is stable in operation. In all cases tests should be made to ensure that the ground can take the load to be imposed (crane plus load lifted). Top soil should be removed and dumped earth compacted. As a safe base for sleepers, lines of concrete laid along the track have proved particularly effective. In winter these concrete bases must be frostproof.

Crane tracks should not be laid too near excavations or excessive ground pressure will be exerted on the edge, and if the side falls in the crane may collapse. The distance between the track and the excavation will depend on the type of crane, the maximum wheel pressure and the nature of the ground. The distance should be calculated from the bottom edge of the excavation. If necessary, bases with a suitable bearing capacity should be put down to support the sleepers. If wooden sleepers are used they should have a cross-section of at least 16 cm by 20 cm. The length of the sleepers should be a quarter more than the track gauge. The spacing of sleepers is determined by the type and size of crane and the bearing capacity of the ground. The necessary guidance should be given in the manufacturer's erection instructions. Half sleepers should only be used for tracks over 4 m wide, and should be 2 m long. At track ends double sleepers should be laid. It is advisable to lay sleepers on a gravel or rubble bed.

If wooden sleepers are used the rails should be laid on sole plates fastened by screws or equivalent means to the sleepers. Only rails with the profile specified by the manufacturer should be used. The track gauge should be maintained in all circumstances. On curves the outer rail should not be raised. To prevent the crane from overrunning the ends of the tracks securely fastened buffers or spring buffers should be provided not less than 1 m from the ends. They should be rigidly connected to the rails or the substructure and should not displace each other. Track-end limit switches must be so arranged that the crane is stopped when necessary at least 1 m from the end. The stop for the limit switch should, if necessary, be carried to the buffers so that the stop cannot be overrun and the limit switch closed again.

The crane track should be as level as possible in both longitudinal and transverse directions. For tower slewing cranes with a jib or with ballast at the top of the tower the track should not be more than 1/4 per cent of the gauge off the horizontal, and for cranes with ballast at the bottom of the tower, not more than 1/2 per cent, in either of the two directions.

Special attention should be paid to the clearances between the track and fixed objects in the vicinity. To prevent crushing it should be ensured that even the extreme points of the crane have a clearance of at least 50 cm. In this connection allowance should be made not only for neighbouring structures but also for future scaffolds and storeplaces.

Qualifications and training of tower slewing crane operators and load slingers

Because of the variety of the technical equipment of tower slewing cranes the qualifications and training of the operators should satisfy strict requirements. Also not every labourer is suitable for slinging loads. Tower slewing cranes should only be operated by reliable persons over 18 years of age who are physically fit, well trained and familiar with the mechanical and electrical installations. Sometimes a medical examination will be required to establish physical fitness and in particular good eyesight, reaction capacity, etc. Coaching of persons in the erection and operation of cranes by the manufacturer's erectors is no substitute for the necessary instruction and training in a regular training course. Before he takes such a course the future crane operator should work for some weeks as an operator's assistant to acquire general experience. The course should equip him with theoretical and practical knowledge relating to the operation and maintenance of cranes, their electrical and mechanical installations, and the relevant safety regulations. At the end of the course he should pass a test on these matters.

Slingers should be thoroughly instructed in their tasks and duties and trained to estimate loads.

Installation, operation and maintenance of tower slewing cranes

Tower slewing cranes should be erected and taken down under competent direction and supervision and in conformity with the manufacturer's instructions. These instructions should always be available at the place of use, and if necessary translated into the language of the person responsible for the work. If live conductors cross the crane's operating area, it should be so placed and erected that no part of it can come dangerously close to them.

When the ballast is being placed the weight given by the manufacturer should be strictly adhered to. If necessary the specific weight of the ballast should be ascertained in advance.

Before the crane is taken into service the operator should examine the track and verify that it is in a safe condition. He should also see that the crane itself is in a safe condition, have the driving mechanisms, running gear, pulleys, ropes, etc., adequately lubricated, and test the overload preventer, limit switches and brakes. Only permissible loads should be slung. If defects are noticed in testing or in operation, the operator should record them in a crane log book to be kept at the workplace, and report them to the competent supervisor, or at the change of shifts to his relief. If a dangerous defect is noticed, he should suspend operation. Failure of a brake or a limit switch, or rope damage due to a rope coming off the drum or formation of knots and loops constitute serious defects.

The crane should be controlled only by the appliances provided for the purpose and it should be so controlled that it operates smoothly. It is wrong to stop travelling and slewing motions by reversing the power. If an irregularity occurs while a load is being moved the crane switch should be immediately opened. Before the operator leaves his stand the crane should be switched off after the load has been set down and the hook raised. When a shift begins after a break, and also when otherwise needed, the operator should give a warning signal. The warning device should not be used to give other signals as for breaks, knocking off, etc.

The crane operator should be able to watch the load, or, if there is none, the lifting gear, throughout all movements of the crane. If he cannot see the loading and unloading points from his stand communication should be ensured by the appointment of signallers who should be so placed that they can see both the load and the operator. Only reliable persons who are familiar with the code of signals should be appointed as signallers. For signalling the internationally recognised hand signals should be used. Loads should only be moved on a signal from the signaller or slinger after it has been made certain that the movement will not cause any danger. If the loading and unloading points cannot be seen, remote control may be used. Two-way talking systems with the main instrument at the operator's stand are also possible. The operator must have properly understood the slinger's signal before he raises the load. He is responsible for correct slinging and should not move overloaded containers, badly made bundles, etc. Loads that have jammed should not be wrenched free. The use of tower slewing cranes for inclined lifting or dragging of loads along the ground or to move any kind of vehicle should not be allowed. Sideways lifting can be avoided at scaffolds and buildings when landing stages are built but they should be correctly calculated so as to be sufficiently resistant to stresses. If grabs are used as lifting gear only loose material should be lifted. The crane operator should be informed before any work is done in the travelling and slewing areas of the crane. As far as possible loads should not be moved over persons. No person should be carried on the carriage of a tower slewing crane, or on the load or the lifting gear. If persons are carried in baskets from which work is to be done the express permission of the competent authority should have been obtained.

No unauthorised person should climb on to the crane. At the close of work for the day and before prolonged interruptions in operations the carriage of track-mounted slewing cranes should be blocked by wind-protection anchorages. The jib should be lowered to the maximum radius and turned into the direction of the wind, but should not be blocked. If however it can be blown against a building or a scaffold, it should be lowered and anchored to a fixed structure in a manner that relieves it of tensile and compressive stress. Tower slewing cranes should be stopped in good time at the approach of a storm and the same measures taken as at the close of day and before prolonged breaks.

When goods are stacked near the track the prescribed clearance of 50 cm should be maintained. Maintenance and servicing operations should only be undertaken by persons with the necessary qualifications. For these operations the electrical installation should be switched off and disconnected from the mains, and precautions should be taken to prevent the current from being switched on again by error or without authorisation.

Testing

Tower slewing cranes should be tested by a specialist before they are first taken into use and thereafter at regular intervals as required by the competent authority. They should also be tested after any structural alteration. The specialist should be recognised by the competent authority. Tests should also be made by a competent person after every erection and otherwise when required but at least once a year. Only mechanical engineers and foremen mechanics of the undertaking, or some other undertaking, especially the manufacturer's, can be considered to be competent persons. For every tower slewing crane there should be a test register and a log book in which the results of tests should be recorded by the person making them.

Building hoists

The term "building hoist" covers both goods hoists in which only goods such as building materials are conveyed and the goods-and-passenger hoists used in the erection of very tall buildings. Naturally the safety requirements for goods-and-passenger hoists are far more comprehensive than those for goods hoists, although the special conditions at building sites and in building operations make some departures necessary from the technical requirements applicable to built-in passenger lifts.

The accident risks in the operation of building hoists are substantially the same as with hoists in general but there are also special risks characteristic of building operations; for instance, the risks of persons falling off the top landing if the hoist platform travels while persons are engaged in loading and unloading, persons being caught by the platform while leaning out into an insufficiently protected shaft, and improperly secured goods shifting on the platform or falling from it.

2.3.6 Goods hoists

(a) Construction generalities

As already indicated goods hoists are to be used only for building materials and the like and not for persons. There are two types, one anchored to a building or a scaffold and the other independent. In both types a crab runs with a swivelling or non-swivelling platform or a skip in one or more vertical or nearly vertical rigid guides that are fastened to a mast made of steel tubes, angle steel or steel latticework. The crab is suspended from a wire rope that runs over top and bottom pulleys to the winch. The winch may have a continuous (one-way) drive or a reversing drive. On independent hoists, which are usually mobile so that they can be easily moved, the winch is built into the hoist frame. Anchored hoists are placed at a suitable distance from the winch.

(b) Soundness and stability requirements

Building hoists and their accessories must satisfy the general requirement to work reliably in normal conditions, and not to entail any risk for the user or the neighbourhood.

Calculations for the steel structure of building hoists must take both the main and the subsidiary forces into account. The main forces are:

Permanent stresses, including external forces acting consistently on the structural parts and causing unvarying strains.

Operating stresses consisting of the load, the dead weight, and the weight of the crab and the platform.

Inertial stresses set up by starting and braking the load and the dead weight.

The subsidiary forces result from wind loading.

To make allowance for the conditions, including shocks, occurring in building-hoist operation, all the normal forces, transverse forces and movements set up by the load and arising in the structural parts should be multiplied by a factor that will depend on the nature of the movement of the platform and on the way the load rests on it.

The resistance of a building hoist to overturning should be determined by a stability calculation providing for the following cases:

hoist operating in still air;

hoist operating in wind;

hoist idle in storm.

(c) _Construction and equipment_

Hoist plates

A metal plate should be affixed at a conspicuous place on the crab giving:

the manufacturer's name or mark;

the maximum permissible load on the platform in kg;

the length and width of the platform in cm.

On the hoist mast at about eye level a metal plate should be affixed giving:

the manufacturer's name or mark;

the maximum permissible load on the platform in kg;

the maximum permissible lifting speed of the platform in m/min.

On independent hoists the height below which the mast need not be anchored should be indicated.

For plates on the hoist winch see section 2.3.1(a).

At the bottom landing of the hoist there should be a notice with lettering at least 8 cm high worded as follows:

Caution! Hoist

Lifting capacity ... kg

Carriage of persons prohibited

Winches, wire ropes, pulleys

The winches, wire ropes and pulleys of building hoists are generally subject to the requirements set out in section 2.3.1 on winches for hoists.

The lifting speed should not exceed 60 m/min. Clutch winches should be equipped with a speed governor that so limits the speed of descent of the platform when carrying the maximum permissible load that it cannot exceed the lifting speed by more than 50 per cent, subject to an overriding maximum of 90 m/min. On winches with a reversing drive it should be ensured by the motor that the platform under the maximum permissible load cannot increase the speed of descent by more than 10 per cent of the rated speed. The controls for raising and lowering should be close together. It should not be possible to operate the winch from the landings on the different storeys, but at each storey there should be means of preventing the platform from descending. The device for this purpose should be coupled to a gate. It should ensure that the operator at the winch cannot move the platform while loading and unloading operations at a higher landing are in progress and persons are still on the platform and allow access to the platform only after it has been blocked. This means that the platform cannot move again until closing of the gate has unblocked it.

This blocking device can be dispensed with on building hoists with a swivelling platform if they are so arranged that it is only possible to load, unload or get on the platform if it is resting on the roof or the scaffold floor.

Travelway

The hoist travelway along which the platform or the skip runs must be so constructed that derailment and jamming are avoided as far as possible. If the conveyance jams in the guides additional friction will be generated, and this will lead to rapid wear of the wire rope on the winch because the pull that the winch must exert to overcome the additional friction will be greater than the stress for which the rope was designed.

When the crab is attached to the wire rope by a compression spring to lessen the shocks during operation the coils of the spring should not touch each other under load.

Guides

The guides of building hoists, usually structural steel pieces such as tubes and sections, should be so calculated that they can withstand, without suffering damage, all stresses resulting from the load or from the operation of the arresting mechanism. The guides should extend so far beyond the necessary lifting height that at the top landing a distance of 2 m remains for overtravel. If the platform overruns the top landing owing to inattention on the part of the operator, the overtravel distance serves to prevent the crab from rising to the pulley and so damaging the hoist.

Overrunning the top landing can also be prevented by means of a mechanical or an electrical limit switch.

Mast

According to the type of hoist the guides (steel tubes or sections) are anchored to the building itself, to a scaffold in front of the building, or to a latticework mast. For this purpose the hoist should be equipped with suitable anchoring devices capable of withstanding the stresses that will occur. At stationary hoists the mast should be anchored in the ground by means of a sole plate or a screw anchor set in concrete. Mobile hoists should be equipped with lugs or similar devices to enable the hoist to keep vertical on uneven ground and help to relieve the wheels. Whether a pile of sleepers will be needed will depend on the bearing capacity of the ground on which the hoist is erected.

Platform

Building hoists with pivoting or swivelling platforms should be equipped with means that effectively prevent the platform from turning on the mast or in the guides while it is being raised or lowered. The devices for this purpose should be so arranged that either they are automatically released at the landing or can be actuated without the operator having to lean over the gate. Since the load on the platform must be prevented from coming into contact with the guides, a tight screen should be fitted that reaches to the top of the crab.

There should be depressible catches or other suitable devices at the end of the rails so that the wagons can be secured against running off the platform as it is raised and lowered. Other loads also must be prevented from falling off the platform: various devices can be used, their form depending on the nature of the load. If the load is in trucks hoops can be used to prevent them from running off. This protection can be supplemented by angle-iron stops fitted on the floor of the platform.

If the load consists of stone placed directly on the platform curbs should enclose the platform on all sides. They are usually so made that they can be removed so that loads projecting over the edge of the platform can be conveyed. Such loads should be suitably supported, when, for example, sheet material must not be subjected to bending stress. The maximum permissible length will depend upon the size of the space in which the platform can turn and the size of the landing. For dumping skips there should be a special device to prevent the skip from dumping while it is being conveyed or while it is at any place other than the landing.

Special safety devices

Arrestor gear. If the rope breaks the crab and the platform will fall with the likelihood of damage and injuries and every building hoist should therefore be equipped with arrestor gear which will act immediately if the rope breaks. There are various types of arrestor; the appropriate type will be determined by the nature and number of the guides. In no case should their efficiency be impaired by the load on the platform.

Protection of the travelway and the landings. If the travelway of the conveyance is in a public thoroughfare or a working area and is therefore accessible, it should be protected by a solid enclosure 2 m high to prevent people from leaning over it. The enclosure may consist of wire netting with a mesh of not more than 5 cm and a wire diameter of not less than 2 mm, or of a barrier placed at least 50 cm from the platform travelway. Where direct access to the platform is necessary at landings, doors or gates with interlocks should be provided. Interlocks should ensure that a door or gate cannot be opened when the platform is not at the landing, and the platform cannot be moved by the winch operator while the door or gate is not interlocked.

Access to machinery. Machinery that requires regular examination should be safely and easily accessible. Lubricating points should be clearly indicated and safely accessible.

Signalling installations. For safe hoist operation there must be means of communication between the operator's stand and the landings. For signalling, bells, gongs, lights, telephones and other means may be used. The code of signals, acoustic or light, should be displayed both at the operator's stand and at the landings. Hoist operators and workers at landings should be perfectly familiar with it.

Erection of building hoists

Building hoists for goods should be erected only under competent direction and in accordance with the manufacturer's instructions. The position of the winch should be chosen with due regard to the necessary angle of inclination of the rope so that the operator from his stand can see as many landings as possible. The mast should be anchored to the building or the scaffold only with the appliances furnished by the manufacturer for the purpose. If the hoist is to be anchored to a scaffold this should first be tested for bearing capacity and stability, and if necessary altered and reinforced. The electrical equipment of the hoist should be installed only by electricians (2.2).

The operator's stand and the landings should be equipped with the prescribed safety devices.

(d) **Tests**

When a hoist is installed for the first time and before it is taken into use it should be tested by a specialist. In addition it should be tested by a specialist at least once a year or at the regular intervals prescribed by the competent authority.

In the tests special attention should be paid to the riveted, bolted and welded joins of the steel construction and the wear of the wire ropes. The working life of the ropes will be determined by the requirements in section 2.3.1(c).

If the tests reveal defective parts they should be removed and if they seriously impair the safety of the hoist it should not be operated until they have been effectively rectified.

Tests should be made with a load 25 per cent in excess of the permissible working load, and all movements - raising, lowering and slewing - should be made at full speed. They should fix the maximum lowering speed of the platform when carrying the maximum permissible load. A hoist should never pass a test if it has any visible defects such as deformations and broken parts. The test must show that the hoist can be operated safely. The efficacy of the arrestor gear should be tested under the maximum permissible load but to avoid unnecessary damage this test can be made with the platform resting in a low position.

Before beginning the day's work the operator should satisfy himself that all equipment is in good working order.

(e)   **Operation of building hoists**

The operation of building hoists should be subject to the requirements applying to the operation of winches as described in section 2.3.1(h). Hoist operators and landing attendants should be familiar with the operating rules and other relevant instructions. No load heavier than the permissible maximum should be moved. The load should be placed properly on the platform and secured against rolling and falling off. It should not project into the travelway.

Leaning into the travelway or remaining at the bottom of it or under the platform or the skip should be prohibited.

Persons should not be conveyed by the hoist.

2.3.7 **Passenger and goods hoists**

Building hoists for both passengers and goods are a fairly recent development originating in the need to convey workers safely and quickly to and from workplaces when tall buildings are being erected. So far no country has issued special regulations for these hoists, and they are still subject to the regulations for built-in passenger lifts. Owing to the constantly changing conditions at building sites, the regulations cannot be fully applied and the general practice has been to make exemptions from some of the provisions where found necessary.

Leaving out of account differences in regulations from one country to another, minimum requirements for the conveyance of passengers on building hoists may be summarised as follows.

For the drive both drum winches and traction sheaves may be used. Drum winches should have no counterweight. Drums should have grooves for the rope and only one layer of rope should be wound on the drum. The diameter of sheaves and pulleys should be at least 40 times, and the diameter of drums at least 35 times the diameter of the rope. There should be at least 2, and with sheave drives at least 3, mutually independent ropes, with a rope safety factor of 14. The cage and the counterwieight should run in rigid guides. The hoists should be equipped with safety gear that acts when the speed of the hoist reaches 1.4 times the maximum permissible operating speed. Whenever the safety gear acts a safety switch should cut off the power.

The hoist should be equipped with a limit switch in the main circuit that cuts off the current on all poles. The switch should prevent the cage moving in either direction. At the bottom of the shaft there should be buffers that brake and stop the cage and the counterwieight if the terminal stops are overrun. The buffers may be fitted on the cage and the counterweight.

The drive should be protected from the weather, and it should be possible to bar access to it.

The fastenings of cage doors or gates should be controlled by a safety switch which prevents any movement of the cage unless the doors or gates are closed.

At the bottom landing the travelway should be enclosed by a transparent method. If slats are used they should be spaced not more than 5 cm apart. Shaft entrances should have transparent doors joined directly to the shaft enclosure. Every shaft door should have an interlock that can only be actuated from the shaft. There should also be an interlock contact that prevents the cage from moving so long as any door is not interlocked, and stops the cage as soon as a door ceases to be interlocked.

Sides of shafts that do not give access to the cage should be fenced to a height of at least 2 m above each landing.

Passengers should only be moved by means of controls inside the cage. Goods may be moved by controls at the operator's stand. External controls at landings are not permissible. In any case if the power fails, controls should return to the zero position. If the cage is controlled from the operator's stand, there should be an indicator showing its position at any moment.

Slinging and lifting gear

A relatively large share of hoist accidents at building operations is contributed by inadequate slinging and lifting gear and improper slinging of loads. Safety in hoisting operations depends greatly on the use of sound and suitable slinging and lifting gear and the secure slinging of loads.

2.3.8 Slinging gear

In hoisting operations the slinging gear comprises ropes and chains.

(a) <u>Chains</u>

For lifting and fastening loads tested chains should be used, that is chains that meet certain requirements as regards material, shape, manufacture, handling and care, and have been tested at the manufacturer's. There should be a works' or an expert's certificate concerning the test. Tested chains should only be manufactured in works authorised for the purpose and subject to supervision. The chains manufactured in these works can be identified by a test stamp.

Chains should not be loaded beyond the permissible maximum weight fixed for them. In hoisting operations they should not be knotted or led over sharp edges or stretched so that they are damaged. Damage must always be reckoned on when the links are subjected to bending stresses, and consequently twisted chains should be untwisted before they are used. Damaged chains should not be used after makeshift repairs. If it is known or suspected that a chain has been overloaded it should be immediately withdrawn from use in lifting operations; this applies also to chains that are visibly damaged. They should not be used again until the damage has been repaired; this will entail technical treatment which can only be given in recognised works authorised to give a test stamp.

A chain should be taken out of use when the whole of it or one link has stretched 5 per cent or more, or the whole or a part of it is impaired by fatigue, or the original link thickness has diminished by more than 10 per cent in any place. Chains need frequent inspection and testing, which should be carried out by competent persons.

If chains are frequently used for the maximum load or are exposed to heat or chemical action they should be inspected every six months. If they are seldom used with the maximum load an annual inspection will do. However, at building operations, chains should be inspected whenever necessary and whenever they are returned to store; before inspection they should be cleansed of oil and dirt.

At every inspection a chain should be examined for external defects, deformation, cracks and wear.

After every second inspection chains should be subjected to a test load, which, as a rule, will be twice the maximum permissible load, but in the case of annealed chains will only be one-and-a-half times that load. If inspection during the test discovers links with cracks or other defects, they should be repaired, and the chain again subjected to the test load and inspected link by link.

Only competent persons should change or insert links, rings and hooks and anneal chains. Maintenance and alteration of special types of chains - case-hardened normalised chains, annealed or high-tensile chains - should only be undertaken by works that possess the necessary experience and equipment. The results of all inspections, load tests, alterations and other treatment of chains should be recorded in a test register or on a chain card to be kept by the contractor.

(b) Ropes

Ropes may be wire ropes or fibre ropes, but all as regards both material and manufacture should conform with recognised rules, which are usually laid down in standards. They should bear durable identification particulars including the maker's name, and in the case of wire ropes the breaking strength of the individual wires, and in the case of fibre ropes the breaking strength of the yarn. The end attachments of ropes should be made in conformity with technical principles. The diamter, type and make of ropes should be suited to the purpose for which they are to be used. Only ropes in one piece should be used for lifting (on winches). It is not permissible to impose a load in excess of the maximum permissible load on a rope, to lead it over sharp edges, or to attach loads directly to it for lifting. Fibre ropes should be stored in dry well-ventilated places, and protected from acids, alkalis and other corrosive substances.

Before they are first taken into use and whenever necessary afterwards, ropes should be inspected by a competent person. The user should constantly watch ropes. If a rope is found to be in a hazardous condition it should not be used again. A wire rope should be considered to be in this condition if a strand is broken, or there are bulges, signs of crushing, sharp kinks, internal or external rusting or many broken wires.

A fibre rope should be considered to be in a dangerous condition if it shows crushing, unravelling, broken strands, large numbers of yarn breakages, or damage due to damp storage or corrosives. The appearance of powdered hemp when the rope is coiled is also a suspicious sign.

(c) Lifting gear

By lifting gear is meant sling chains, chain slings, sling ropes, rope slings, skips, platforms, grabs, tongs, claws, lifting beams, etc., very many types of which are used for transporting building materials and structures (Fig. 10).

When chains and ropes form part of lifting gear the requirements of sections 2.3.8(a) and (b) apply. Like other chains, sling chains and chain slings should only be made by competent firms. On grabs, skips, tongs, claws, clamps and lifting beams the maker's name should be shown, or should be otherwise known. On all lifting gear the maximum permissible load should be clearly and durably indicated; it should never be exceeded. Platforms for loading stones should have rims on all sides unless safety baskets are used. These should be so fastened to the platform that they cannot inadvertently be detached. Pulley blocks should be so made and installed that no one can be caught in the pulley or between lifting gear and the pulley. Loads should be so slung or stacked on the lifting gear that they cannot fall down. If this is not possible suitable containers should be used; they should not be loaded above the rim.

Rings, links, etc. of lifting gear should turn freely on the hook. Lifting gear should not be thrown down as that might damage it. For sling chains and ropes a sufficient number of loading tables, durable and easily legible, should be displayed at suitable

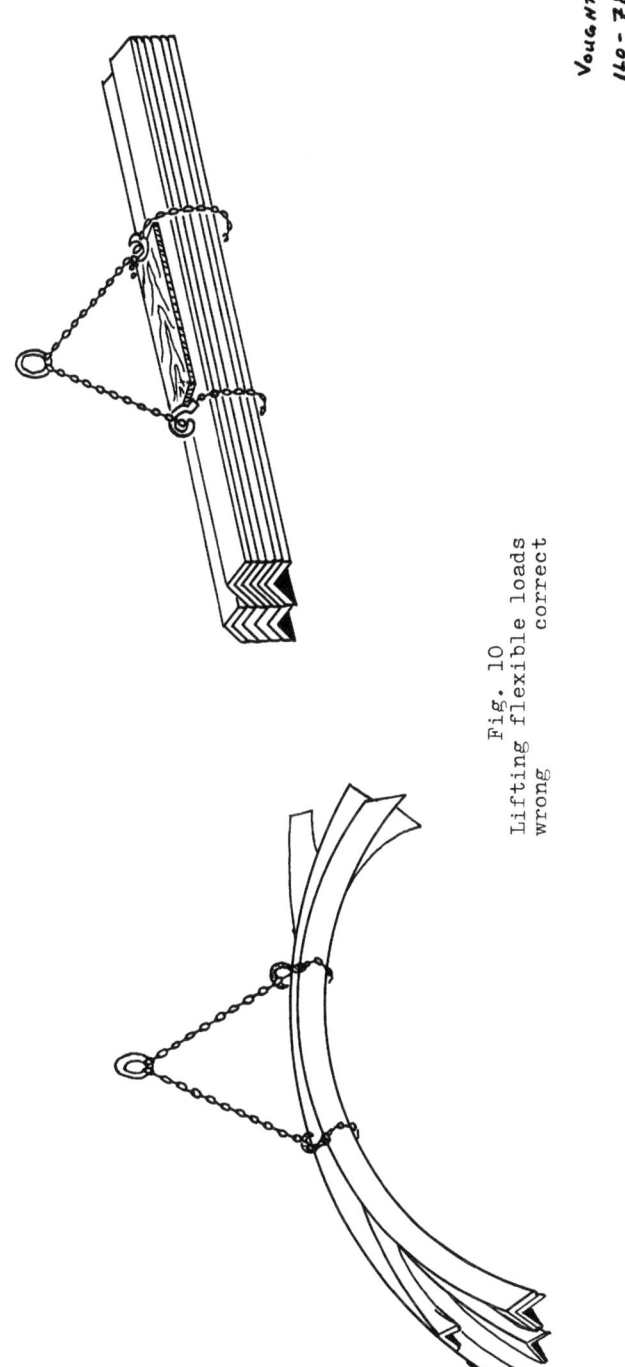

Fig. 10
Lifting flexible loads
wrong   correct

places. These tables in a handy form should also be given to crane drivers and slingers. The maximum permissible load at different angles of the legs of chain and rope slings and attachments should be calculated. Angles exceeding 120° should be avoided. If chain and rope slings consist of more than two legs, it should not be assumed that all are bearing the load unless it has been made certain that the load is equally distributed among them. If this is not so then only two legs should be assumed to be bearing the load. Unloaded sling chains and ropes should be hung on the crane hook, which should be raised.

Lifting hooks should be safety hooks, because a load that is not guided might strike against the building or the scaffold and fall off. With hand-hauled ropes as when buckets are lifted, snap hooks can be used as safety hooks, and with power-driven lifting appliances, hooks with latches.

Like chains and ropes, lifting gear should be tested before being taken into use for the first time and thereafter as required by a competent person: the results of the tests should be recorded in a register.

Conveyance of persons on lifting appliances at building sites

2.3.9 General

As mentioned already the conveyance of persons on lifting appliances at building sites, that is riding on the lifting gear or the load, is generally prohibited. Where in special circumstances such conveyance is desirable or even necessary for accident prevention purposes, application should be made to the labour protection authority for an exemption, which should be notified in writing. The competent authority at the same time should set out the conditions in which persons may be conveyed.

Conveyance of persons on lifting appliances at building sites may be necessary when workers have to be taken to and from their workplaces in a relatively simple way during the erection of tall chimneys, television towers, bridge pillars, tall silos, etc. The same is true when workplaces are in deep shafts. The construction and climbing of extensive ladderways in such building operations is undoubtedly fraught with greater danger than the conveyance of persons on lifting appliances provided that the necessary precautions are taken.

Conveyance of persons on lifting appliances may also be permissible when work has to be done from a cage suspended from the appliance, as may happen in steel erection or in assembling prefabricated reinforced concrete units. In these cases too, the provision of a safe working place in an exposed position may require measures that conceal considerable risks.

2.3.10 Construction and equipment

Authorisation for the conveyance of persons on lifting appliances at building sites can only be granted subject to conditions that will exclude the risks. These conditions will be concerned with both the construction and equipment of the lifting appliance and the construction and equipment of the cage that will be used with it. Not every winch is suitable for the conveyance of persons. Those used for the purpose should be driven by electricity and have sequence switching. When the controls are released they should automatically return to zero. If the power fails or the motor is switched off the brake should be applied and the load securely held. There should be an arrangement whereby if the lifting motor fails the cage can be lowered to its point of departure. This arrangement should be effectively protected by a lock against unauthorised use. Out of doors the lifting speed should not exceed 0.85 m/sec, and indoors 1.5 m/sec. The highest working position of the cage should be fixed by a limit switch. Allowance should be made for inching. It should be possible to run backwards.

For lifting, a non-twisting wire rope with low internal stress at least 8 mm in diameter should be used. The breaking strength of the wires should not be less than 140 kgf/mm^2 and not more than 160 kgf/mm^2. A works certificate should accompany the rope.

The diameter of the winch drum and of the pulley should be at least 24 times that of the rope but in the interests of safety every effort should be made to use a larger diameter.

The distance of the winch from the bottom pulley should be at least 10 times the length of the drum. If this is not possible for reasons of space, and more than two layers of rope are wound on the drum, a rope coiling device must be used.

When persons are conveyed a special cage is necessary and not more than two persons should be carried on it: the maximum permissible load should be marked on it, clearly and durably. The entire construction of the cage should be covered by a certificate of soundness based on tested calculations. It should be suspended from a twist preventer. The hook should be of the safety type.

The cage should be enclosed on all sides to its full height, by wire netting for example, and have a solid roof and a door with a reliable lock. The lock should be equipped with a device that effectively prevents the door from opening inadvertently during travel. The cage should have a double floor with an intermediate layer of elastic material.

If work is to be done from the cage it should have a solid enclosure at least 1 m high. The soundness of the construction should be attested by a certificate. An automatic landing device, such as a flap on the scaffold, or the lowering of a sufficiently strong door leaf should prevent the cage from being lowered inadvertently while it is being entered or left. Openings for passage through ceilings, platforms, scaffolds, etc., should be so arranged as to ensure unhindered passage. For communication between the winch operator's stand and the landings an electrical signalling system or a telephone should be installed.

Testing

Equipment for conveying persons should be thoroughly tested at the building site by a competent person before it is first taken into use and after any lengthy interruption in service. Before it is first taken into use it should be tested under 1.25 times the maximum permissible load and in all directions of travel. Any defects discovered should be remedied before it is taken into use. The equipment should also be thoroughly tested at least once a week, and be inspected externally every day.

Before persons are conveyed on a tower slewing crane it should be examined by the competent specialist. The measures then decided upon should be properly carried out before passenger operations begin.

Operation

A responsible person should be appointed by name at the building site to ensure that the rules for conveying persons are strictly observed: he should be handed the relevant regulations and a copy of the official exemption. Passenger equipment should only be operated and serviced by reliable persons at least 21 years of age who are familiar with it.

The prescribed tests should be carried out by a specialist in the presence of the appointed responsible person and the winch or crane operator.

The cage may only be entered and left when it rests on the floor or another sufficiently firm supporting surface or when the landing device is fully effective. Starting and stopping should be preceded by a clearly perceptible signal. Changes in the signalling arrangements should not be allowed. When the cage approaches a landing its speed should be reduced by stepping down the winch power.

Only the work specified in the exemption should be done from the cage. At the workplace the cage should be secured against swaying. Tools and material carried in the cage should be secured against displacement and overturning. Persons in the cage should be secured by safety belts against falling out. If during operation defects are found in the equipment for passenger conveyance, they should be immediately reported to the responsible person who should take steps to remove them. If necessary, a specialist should be called in, a mechanical engineer or a foreman mechanic for example.

2.4 Mechanical conveyance

Belt conveyors

2.4.1 General survey

Transportable belt conveyors whether mobile or fixed as used in the building industry for moving earth, rubble, additives, stones, etc., are not generally considered to be particularly dangerous appliances perhaps because of the relative slow speed of the belt. However, there are an appreciable number of accidents at these conveyors. A large proportion of them are due to defects and occurrences common to all mechanical equipment on building sites. They include accidents due to kicking back or falling out of the starting handle of internal combustion engines, and the creation of dangerous contact voltages on damaged electrical equipment, a danger to which mobile belts are particularly exposed when they are moved without being disconnected from the mains. There are also the accidents in maintenance work on moving equipment, in unloading belts on to vehicles, in shifting belts without lowering the ends so that they touch overhead electric lines, and in unauthorised climbing on belts to reach a higher workplace with the result that persons fall off.

There are also accidents peculiar to belt conveyors:

- accidents that occur in cleaning drums and the belt, or in removing stones while the belt is running;

- accidents due to stones or other loads falling off or rolling;

- accidents in removing stones from the top end when the belt is used instead of a hoist to transport stones to a higher level;

- accidents due to sitting on the end to achieve a better distribution of weight when a belt is being shifted by hand.

2.4.2 Safety requirements

The prevention of accidents at belt conveyors calls for the same precautions as described for mechanical equipment. In addition the following requirements should be met.

(a)   Power transmission

With many belts the motive power is tranmitted by V belts from the motor to the driving drum. These drives are usually enclosed but frequently only in front or at the sides. They should be effectively enclosed on all sides, and if only for the sake of cleanliness the enclosure should be of sheet metal.

(b)   Drum nip points

These points constitute the greatest danger at belt conveyors, and consequently their arrangement deserves the closest attention. Accidents at these points are almost always very serious because a person who is drawn in between the belt and a drum often loses a limb, if not his life. If a tool such as a shovel is caught at these points the most serious injuries may be caused by a blow or by crushing. Nip points should therefore always be enclosed. Enclosures may be confined to in-running points where the belt changes direction. This is always the case with driving drums, tail drums and tensioning drums. Special attention should be paid to nip points of belts that can run in two directions.

Protection should always extend over the whole width of the drum. It can usefully consist of a piece of wood or angle steel and be so fastened that it cannot be merely lifted off. Lateral enclosure of nip points is not enough especially if it can be removed or otherwise displaced. The sheet metal guard at the side of the nip points between a drum and the belt should be so fashioned that the tensioning drum can be adjusted without removing it.

(c) **Arrangements for safe cleaning of the belt and the drums**

It is an unvariable rule of accident prevention that machinery should be cleaned only when it is stopped. At building operations the loads and the weather may make the belt dirty very quickly and fragments of material may adhere both to the belt and the drums. Frequent cleaning is necessary and work would be greatly impeded if the conveyor had to be stopped every time. Cleaning with a long-handled appliance is not completely safe but scrapers should be fitted which can be safely operated from outside the enclosure even when the belt is running. The arrangement must be such that the scrapings do not fall on the return side of the belt. If the underside of the belt is completely enclosed, on the enclosure in front of the reversing drum there should be a diagonal deflector that conveys falling material to the outside.

(d) **Special precautions with mobile belt conveyors**

Mobile belt conveyors should be firmly set up. On steeply sloping ground they should be secured against sliding down; unless they are equipped with a special brake for the travelling motion they should be suitably blocked. Mobile belt conveyors with an adjustable transport level should only be suspended from a rope if an irremovable safety device is provided. On belt conveyors with oil-filled hydraulic equipment for raising and lowering the jib the lowering speed should be limited by arrestor devices, for otherwise if the regulating screw of this equipment were wrongly handled the jib might drop too fast and too far, to the danger of any persons underneath. When belt conveyors are moved their stability should be ensured. They should only travel when in their lowest position, and should be stopped before they move. If the drive is electric, the conveyor lead should be disconnected from the mains. While a belt is being transported, shifted or turned no one should be on the belt or hang on it as a counterweight. If the loading mechanism cannot be raised, the belt end should be pulled down by ropes.

Pneumatic conveyors

Compressed air is used at building operations both as a source of pressure, as in plastering machines, and as a means of transporting concrete. In both cases the equipment consists of a mixing chamber, an air receiver and piping. The air delivered by a compressor is first stored in an air receiver and then led to a mixing chamber. When the trigger is pressed the compressed air enters the mixing chamber and expels the contents into the piping.

2.4.3 **Plastering machines**

(a) **General**

On these machines pumps force mortar through pipes or hoses by compressed air on to the wall to be plastered, and accordingly the majority of the accidents occurring on them are caused by spurting mortar. The accidents often cause more or less severe eye injuries. Particularly severe injuries occur when clogged mortar pipes are being cleared if the pressure in the system has not been cut off, through forgetfulness or neglect.

If accidents are to be avoided the following safety requirements concerning the construction and use of the machines should be met.

(b) **Machines**

The air receivers and pressure equalising appliances of plastering machines should satisfy the technical safety requirements for pressure vessels (2.8) and the equipment furnishing the compressed-air supply, those for compressors (2.8). To prevent overpressures in the machine and the piping there should be devices that automatically switch the machine off at a certain pressure. Various arrangements can be employed depending on the type of drive. With internal combustion engines a clutch like those on motor vehicles can be used that automatically stops the engine when a certain torque, which can be varied, is reached. With electric drives, when there is an overload the power is switched off by a contactor control with a protective motor relay. With hydraulic machines, at a certain pressure an overload valve short-circuits the oil piping, thus disconnecting the machine and stopping the pumps. The mortar pump can be stopped while operations continue by means of a remote-controlled air regulator operated from the spray gun. Closing the remote control or air stopcock at the spray gun will raise the pressure in the compressor, and then the remote control will act to disengage the machine.

(c) **Mortar piping**

To prevent inadvertent disconnection of lengths of piping or hose, their connecting pieces should be equipped with safety fastenings. Pipes and hoses should not be too long, and should be laid along the shortest route to the workplace. When hoses are being laid care should be taken to avoid sharp bends, kinks and unnecessary turns. Rising pipes and hoses should be firmly secured if only because their weight is considerably increased while mortar is being pumped. Connecting pieces should be thoroughly cleaned before pipes or hoses are coupled up; otherwise leaks and blockages may occur. Hoses should be protected against sharp points and edges as well as against oil and petrol.

(d) **Spray gun**

Special attention should be paid to the tightness of the air cock or trigger of the spray gun. If it is too loose, it can open unexpectedly and the mortar spray may seriously injure people.

(e) **Operating precautions**

It is particularly important to comply strictly with the operating instructions. Everything possible should be done to avoid blockages in the piping and so eliminate one of the main sources of accident. Blockages may result from inadequate lubrication of the piping, dirty and therefore leaking connections, penetration of foreign bodies into the pump piping or spray gun, seizing of a worn-down ball in the valve seating, mortar that is not easily pumpable, or sticky or separates out too easily, such as gypsum mortar, thin sharp mortar and pure cement mortar, kinks and bends in the piping, and changes in the cross-section of the piping.

(f) **Precautions in the event of blockages**

If a blockage does occur and pipe or hose connections or other mechanical parts have to be disconnected or dismantled, the work should only be done after the machine has been stopped and the pressure released.

(g) **Personal protective equipment**

To avoid eye injuries it is absolutely imperative that both machine operators and plasterers should wear tight-fitting protective goggles. The fact that goggles quickly get dirty and so must be cleaned frequently is no reason for not wearing them. When dealing with blockages it is also advisable to look away from the equipment, and no unnecessary persons should be near.

Other personal protective equipment that has rendered good service includes rubber boots and gloves for sprayers. See also section 4.

2.4.4 Pneumatic concrete transport

(a) **Measures for counteracting recoil and centrifugal forces in the piping**

As regards the technical characteristics of the plant, the recommendations made in section 2.4.3 are applicable in the main. From the safety standpoint particular importance attaches to the piping. Of the forces likely to cause accidents special mention should be made of the recoil of the moving masses and the centrifugal forces in the bends of the piping. These forces are incomparably greater than those present in plastering machines. The magnitude of the forces can only be effectively influenced by the speed at which the material moves through the piping. Opposed to the forces in the piping are the forces at the supporting points of the piping. The force of the recoil can be absorbed by a receptacle fitted at the end of the piping that divides the issuing stream of concrete and destroys the kinetic energy in it by frictional forces. However, the transverse forces in the bends of the piping can only be dispersed at points of support in the ground or in fixed structural parts. It is hazardous to disperse forces directly through concrete forms. Special measures are needed, for usually concrete forms are not designed to withstand the forces generated in the piping. It is therefore advisable to support the piping and the receptacle independently of the scaffolding and concrete forms. If bends cannot be avoided they should never be at the ends of piping. In no case should the terminal length of piping consist of hose that has been bent to suit conditions on the spot.

(b) <u>Clearing of blockages</u>

The chief sources of disturbances in operation are blockages in the piping, which in most cases are due to deposits, unsuitable composition of the concrete, or inadequate motive pressure. As the operating instructions will indicate, to clear a blockage the control valve should be closed and the pressure in the gun reduced to 1.5 ats. gauge by letting air escape. Then an attempt may be made to clear the blockage by banging the piping. If the attempt is not successful all the air must be let out so that the gun can be opened. Then the piping, which will not be under any pressure, can be disconnected, dismantled and cleaned length by length. It would be extremely rash to try to remove adhering concrete by applying extra pressure.

(c) <u>Requirements concerning operating personnel</u>

Since the smooth and safe operation of pneumatic transport equipment requires considerable intelligence, operators should be suitably qualified persons who have been minutely instructed in their duties and made fully aware of the consequences of any incorrect action.

2.5 <u>Scaffolds</u>

The scaffold is probably the oldest and at the same time the commonest piece of equipment used in building operations. With the development of new types of scaffold the form has undoubtedly changed but the purpose remains the same. As a work scaffold it has to carry not only the workmen and their tools but also the building materials needed for the work; as a protective scaffold it has to protect workers against falls and to prevent material and tools from falling on them; as forms it supports parts of buildings until they are self-supporting as the result of setting of mortar or concrete or assembling of structures.

The technical development of scaffolds has been mainly influenced by the requirements created by the employment of new building methods. These have also helped to make the erection of scaffolds safer. More than the old, the new scaffolds clearly show how important it is to conform to recognised technical rules in scaffold erection.

In the main, causes of scaffold accidents are breaches of recognised technical rules. Instances are: defective or inadequate structure of scaffolds or their parts, use of unsuitable or damaged scaffold material or parts, overloading the scaffold, inadequate bracing, and in the case of scaffolds fastened to the building, above all inadequate and improper anchoring. There are also falls of persons during erection and dismantling of scaffolds, or due to the absence of railings and toeboards when work is being done from scaffolds.

General requirements

Among the different types of scaffold are pole scaffolds, outrigger scaffolds, bracket scaffolds, suspended scaffolds, trestle scaffolds and supporting scaffolds for erection and storage purposes. These last include concrete forms and their supports.

The requirements concerning the stability and structure of scaffolds may be summarised in the following general principles:

- the scaffold and its parts should be so designed that it can perfectly withstand and safely release the vertical and horizontal forces imposed on it in use, either by itself or in conjunction with load-bearing building parts;

- use should only be made of good and suitable scaffold material that has been examined as to its fitness by a competent person. If the material has to conform to certain specifications this should be ensured;

- to enable the scaffold to absorb horizontal forces it should be reinforced in the longitudinal and transverse directions by diagonal bracing. A scaffold that is not independent should be sufficiently and properly anchored to fixed parts of the building;

- connections on scaffolds should so join parts that they can bear the imposed loads;

- the scaffold floor should be suitable for the type of work. It should be tight and so laid that the boards can neither tip up nor fall out;

- scaffold floors over 2 m above the ground, and openings in them, should be fenced with two-rail railings and toeboards;

- scaffold floors that are used should be provided with a safe means of access;

- erection and dismantling should be done under competent direction and supervision;

- scaffolds should not be used before they have been completed. Before they are used, and after lengthy interruptions in the work, they should be examined for stability, and the connections examined for strength;

- scaffolds should not be loaded to an extent exceeding that for which they were planned.

Calculation of scaffolds

The type and specifications of a scaffold will be determined by its purpose. Since the purpose may alter as the work proceeds it is important in planning to decide which requirements the scaffold must satisfy. Only in this way can the most suitable type of scaffold be chosen and subsequent additions and alterations, which are usually troublesome, avoided.

When planning a scaffold it is necessary to consider how many loaded levels should be provided for, whether all levels should be floored, or whether the flooring can be taken from level to level as the work proceeds. Thus the scaffold is to be designed to bear a certain load, and there will be substantial differences between requirements for masonry work and plastering and light maintenance work, for which little material is wanted. In all cases a factor of safety of four times maximum load should be used in the design of scaffolds.

For the most important building operations the following load-bearing capacities may be taken as acceptable:

- for masonry work: uniformly distributed load of 350 kgf/m²;

- for plastering: uniformly distributed load of 250 kgf/m²;

- for maintenance work, painting, etc.: uniformly distributed load of 125 kgf/m².

These capacities are based on the assumption that the load on one level will be as follows:

- on masons' scaffolds: 2 masons, 1 stone or mortar carrier, 1 box of mortar and 150 bricks;

- on plasterers' scaffolds: 2 plasterers, 1 mortar carrier, 2 boxes of mortar;

- on scaffolds for maintenance work requiring little material: 2 workers, tools and materials.

If on individual parts of a scaffold, flooring for instance, there are heavier loads, the following individual loads should be reckoned on:

- on masons' scaffolds: 150 kgf;

- on plasterers' scaffolds: 150 kgf;

- for maintenance work: 75 kgf at intervals of 50 cm.

In this calculation an individual load of 150 kgf represents a stone or mortar carrier, and one of 75 kgf two persons standing close together.

If, because of the method of working or the small scale of the work, it can be ensured that the assumed loads for masons' and plasterers' work will not actually be placed on the scaffold, as an exception lower capacities may be allowed but not less than 100 kgf/m² or two individual loads of 75 kgf/m² each 500 mm apart at the most unfavourable place.

On the other hand, in cases in which the method of working causes the capacities quoted above to be exceeded, the actual loads should be ascertained and the construction of the scaffold and its parts based on them. In order to ensure that in use the scaffold will not be improperly loaded beyond the extent on which its design is based, a notice should be displayed at conspicuous places with wording such as "Maximum load 200 kgf/m²". This is particularly necessary when a scaffold has not been erected by the user but by some other firm, or when it is to be expected that it will be used by a number of contractors for different purposes.

For scaffolds that are only used for protection the loads assumed for light maintenance work with little material are applicable.

Concrete forms should be calculated not only for the weight of the fresh concrete or the masonry and any reinforcement to be placed upon it, but also for a load representing the forces acting in concreting by reason of the dumping of fresh concrete from containers, the workers and the traffic on the forms, etc. The size of this additional load will depend largly on the capacity of the containers. For concrete floors of the dimensions commonly found in house building it is taken to be kgf/m².

In the planning of scaffolds, in addition to the vertical loads horizontal forces, such as wind loading and the forces exerted by lifting appliances, etc., have to be taken into account. In addition account must be taken of horizontal forces generated by unavoidable tilting of supports. For this purpose, in addition to the actual horizontal forces, a force of at least 1/100 of the vertical loads to be borne by the scaffold floor or the forms is assumed to act at the level of the floor or the top of the form.

If the mortar or the concrete is conveyed by piping resting on the scaffold or forms account must be taken in planning the scaffold of the resulting vibration, unless special measures are taken to support the piping in such a way that the strokes of the pump are not transferred to the scaffold.

If a scaffold is designed on the basis of static calculations, a drawing should be made for the site to ensure that the assumptions made in the calculations are respected in the construction.

Constructional details

2.5.1 Flooring

The thickness of scaffold flooring will depend on the loading and the distance between supports. On work scaffolds the floor thickness should be at least 30 mm, and on forms it is usually 24 mm. As regards other minimum thicknesses for certain types of scaffold, see section "types of scaffolds".

It is useful to fasten a corrugated metal band on the ends of floor boards of work and protective scaffolds to prevent splitting. The common practice of nailing an ordinary metal band to protect the ends of boards has not proved satisfactory because the band is easily torn off in the rough conditions of building operations and so may be a source of stumbling and hand injuries.

Floor boards should be laid touching each other and so that they can neither tip up nor fall out. This requirement is satisfied when under every join in the flooring there are two putlogs, or the boards overlap at least 20 cm on a putlog.

The thickness of the flooring can be reduced by fastening the boards with cross pieces to tables so that all the boards share the load, but in some cases the strength should be certified.

On scaffolds with uprights on both sides, the floor should cover all the space between the uprights, and on scaffolds with one row of uprights and one end of the putlogs anchored in the masonry, the floor should be brought as close as possible to the building.

Floor boards should not project more than four times their thickness beyond the last supports. High scaffolds should be secured against being blown down if the weather conditions so require.

2.5.2 Edge protection

All scaffold floors and openings in them, such as those for ladders, should be protected at the edges to prevent falls of persons and materials, tools, etc. This protection will consist of a two-rail railing with a toeboard. If loose material is stored on the scaffold the intermediate rail and the toeboard should be replaced by a hoarding about 60 cm high.

Depending on the type of scaffold, railings will be composed of bars, boards or steel tubes. The dimensions of the upper and lower rails will vary with the spacing of the posts to which they are fastened. For posts spaced 2 m apart, boards with a cross-section of at least 150/300 mm will be required. The top edge of the railing should be 1 m above the scaffold floor, and that of the lower rail, 45 cm. The toeboard should be at least 15 cm high and be secured against tipping over.

If the space between the scaffold floor and the building exceeds 30 cm, edge protection identical with that on the outer side of the scaffold is also required for the inner side. This is also the case when there are openings in the masonry at the level of the scaffold floor.

2.5.3 Bracing

To prevent displacement, over the entire face of the scaffold bracing should be installed beginning at ground level and running at an angle of about 45° to the top. For some types of scaffold cross bracing is usual in an arrangement in which in addition to the spaces between the last two uprights at each end, every second space between two uprights is braced.

On independent scaffolds transverse bracing is also necessary. At all points where it crosses uprights and ledgers the bracing should be fastened. The usual scaffold bars, boards, steel tubes and the like can be used for bracing. The bracing should not be removed until the scaffold is finally dismantled.

2.5.4 Anchorage

In the construction of scaffolds that are not independent, which may usually be called facade scaffolds, proper anchorage is an important element in their stability.

Anchoring serves both to limit the buckling length of the uprights to that planned for the scaffold, and to prevent the scaffold from coming away from the facade and collapsing. A relatively large proportion of known collapses of scaffolds is attributable to faulty anchoring. Factors of decisive importance for the security of the anchoring are the number of anchors, the type of anchor and the method of anchoring. The vertical and horizontal distances between anchors should not exceed 4 m; the anchoring points may be staggered from upright to upright. In this way every upright is anchored even if the uprights are spaced less than 4 m apart. In exceptional cases, if the nature of the facade makes this necessary (e.g. high shop fronts) the lowest anchors may be higher than usual, but not higher than 6 m. Every upright should be anchored at least twice to the building. The scaffold should not extend more than 2 m above the highest anchorage. If the scaffold is used with a protective wall as a protective scaffold for roof work, the uprights should be anchored to the roof construction with due regard to the area of the protective wall exposed to wind pressure.

The anchoring points for scaffolds that are not independent should be so chosen that they do not hamper subsequent building operations and cannot be prematurely removed. For this reason it is not advisable to use window openings for anchoring scaffolds, because when the windows are being placed and the sides of the openings plastered, the anchors are a nuisance and have to be removed. If during the work anchors have to be removed this should only be done if they have been replaced by other anchors. Anchorages should be made as the scaffold rises and not after it has been erected. When a scaffold is dismantled the anchorages of the uprights should remain until the uprights themselves can be taken down.

To achieve an effective anchorage that will absorb tensile and compressive stresses, good anchoring devices are needed that will ensure secure attachment to the building and to the scaffold.

As a rule clamps are used that are driven into the masonry. Expanding plugs are also used; they are placed in the masonry, and devices for fastening the scaffold can be screwed into them. Shooting anchoring devices into the masonry has not proved satisfactory. The anchor should be firmly embedded in the masonry. When hollow bricks or blocks are used care should be taken that the anchors are held tight. As far as possible firm parts of the building should be chosen for anchoring. Difficulties arise with completed facades when owing to the construction (plaster, sheet facing, etc.) it cannot be ascertained where the firm parts, such as concrete floors and pillars, are. Scaffolds should never be anchored to drain pipes, lightning conductors and the like.

If while a building is being erected anchors are provided for the subsequent placing of scaffolds for maintenance work, they should be sufficiently corrosion-proof, and before they are used they should be examined as to their good condition and strength.

2.5.5 Scaffolds for special purposes

(a) Protective scaffolds

If they are to achieve their purpose, protective scaffolds should be so made that persons, materials, tools, etc. cannot fall off them or break through them. Consequently they must satisfy two requirements: first they should be so wide that a person falling does not plunge over the edge, and second they should not be so far below the workplace that a falling person can injure himself by striking against them. Also the scaffold floor should be so strong that a falling load will not break through it.

As a rule the floor of a protective scaffold should not be more than 3 m below the working place, and it should be at least 1.3 m wide.

Almost every type of scaffold - pole, outrigger, bracket, suspended - can serve as a protective scaffold and when it does, in addition to satisfying the two above-mentioned requirements (minimum width and position of floor), it should have a protective wall which as a rule should be 1 m high. If, however, a work scaffold is used as a protective scaffold for roof work, the protective wall should be higher so that a person falling is not carried beyond the scaffold by the slope of the roof. In such cases the scaffold should not be more than 1.5 m below the edge of the roof, for otherwise the protective wall would have to be too high and so offer too large a surface to wind pressure.

(b) Inside scaffolds

Independent scaffolds are used in most cases for work on ceilings in rooms and halls. These scaffolds should be particularly well reinforced, and braced sufficiently against displacement. To establish the dimensions the stresses should sometimes be calculated. In addition to the vertical loads, account must be taken of horizontal forces such as those resulting from the use of a lifting appliance on the scaffold.

The scaffold floor should be close-planked and enclosed on all sides by railings and toeboards. To avoid overloading a scaffold, a table should be provided showing the maximum permissible load. In no circumstances should persons collect at one spot on the scaffold as may happen during inspections or on the ladder during breaks. For access to the scaffold a ladder should be provided. Climbing up and down the scaffold should not be allowed.

(c) Travelling pole scaffolds

When pole scaffolds of any type are built as travelling scaffolds, the wheels, which may be placed directly under the uprights or under a plate, should satisfy certain requirements. The uprights should be so connected to the carriage that they cannot sink or be lifted out. In the construction of travelling scaffolds account must also be taken of the forces exerted on them and individual parts when they are moved by hand or by a lifting appliance. When a scaffold is moved, care should be taken that the application of force to a part does not cause it to be displaced or even to break.

The wheels of travelling scaffolds should be equipped with blocking devices so that a scaffold cannot inadvertently move while anyone is on it.

The use of travelling scaffolds implies that the ground on which they are to move is not uneven and is firm enough to prevent the wheels from sinking in. If necessary steel girders should be laid on the ground to serve as guides for the wheels; these enable the scaffold to move much more easily.

Because of the forces needed to move them travelling scaffolds require particularly rigid three-dimension reinforcement.

Types of scaffold

2.5.6 Pole and similar scaffolds

Pole and similar scaffolds are scaffolds with which the loads on the floors are transmitted through vertical poles, standards, etc., to a firm base such as the ground, a floor or girders. As wooden scaffolds these scaffolds can be made of squared timber, round poles or ladders, and as metal scaffolds, of steel or light-metal tubes, steel or light-metal sections, or frames of steel tubes or steel sections, the connections varying with the different forms. With wooden scaffolds the connections are bolts with screws, clamps, chains, wire ropes, lashings and cords; with steel scaffolds they are couplings or pins. A distinction is drawn between independent and tied scaffolds. Independent scaffolds may be built as travelling scaffolds.

A distinction may also be drawn between outside and inside scaffolds. Depending on their purpose the former have floors ranging from 60 cm to about 1.5 m wide, while with the latter the size of the floor will depend on the scope of the overhead work.

(a) Squared-timber scaffolds

These scaffolds are structures of sawn wood, rarely of half-round or round timber, assembled by bolting. They have many uses. They are in general use for supporting structures and for staging or trestlework. As work scaffolds they are used to support loads of more than 400 kg/m². Squared-timber scaffolds should not be built by semi-skilled workers; the carpentering calls for fully qualified persons. In any case the stresses should be calculated and drawings made. With a view to more economic use, these scaffolds, whether internal or external, are sometimes built as travelling scaffolds.

(b) Round-pole scaffolds

With these scaffolds the uprights, ledgers and putlogs are made of round timber. The poles should be straight and of uniform growth. Before they are used the bark should be removed and the tops cut off to the point where the minimum permissible thickness is reached. For connections, cord, wire rope, chains, lastings, clamps, etc., may be used. Round-pole scaffolds are mainly used as work and protective scaffolds supporting loads between 100 and 400 kg/m².

The scaffolds may have one or two rows of uprights. As masons' or protective scaffolds in rough conditions they usually have one row, that is on the inner side the putlogs rest on the masonry. Where this is not possible because holes cannot be left for them or cannot be made in existing masonry the scaffold should have two rows of uprights, as, for instance, when facings or hollow-brick walls are being built. Two-row scaffolds are also used for plastering.

It is difficult to draw up generally applicable specifications for round-pole scaffolds and their parts. For this the purposes of the scaffolds and local working methods are too varied, as are also the regulations and the minimum dimensions in force in the different countries.

Poles

Poles should be let into the ground from 0.60 m to 1 m and secured against sinking in further by resting them on sole plates. If conditions on the spot make it impossible to sink the foot of the pole it should rest on a board or a beam and be secured against lateral displacement by nailing on slats or knocking in clamps. When poles rest flush on the ground in this way the scaffold needs to be particularly well braced longitudinally, or if it is a two-row scaffold, in both directions.

Where, as at drives and entrances, poles have to be farther apart, the ledgers may have to be reinforced or braced or intermediate uprights used to form a frame.

Ledgers

The ledgers, which support the putlogs on which the flooring rests, are best placed inside the uprights. They should be laid at intervals not exceeding 4 m, or two scaffold levels, and should remain in place until the scaffold is finally dismantled. They should be joined by an upright and the two ends should overlap 1 m and be properly fastened together. Ledgers should never be joined in the space between two uprights, and loads should not be placed on unsupported projecting ends of ledgers.

Putlogs

Putlogs on which the flooring rests directly should be round timber with the cross-section intact; they should never be split lengths. Their soundness should be verified by tapping before they are used. Putlogs that rest on the masonry may have two sides made parallel at the end so that the thickness on the support corresponds to that of a course of masonry and so allows the end to be easily walled in. The length supported in the masonry should be at least 15 cm, or half a brick length. Putlogs that are opposite an opening in the masonry and have no firm support in it or on any other firm part of the building should be secured by girders or beams so that they cannot move.

Connections

Connections for pole scaffolds should satisfy certain requirements as regards both their strength and the means of fastening them, because the coupling that they form must be strong, rigid and immovable. If wire ropes, clamps or cords are used as ties, joints between ledgers and poles should be supported by stays or props. Additional support is not needed when tested chains are used. Cords, which may be made of hemp or jute, should be long enough to go at least five times round a pole. Sisal ropes should not be used on scaffolds because of their brittleness.

Anchoring and bracing

Tied pole scaffolds should be properly anchored to the building. Vertical and horizontal distances between anchors should not exceed 6 m. The anchors should be staggered and be so fashioned that they can transmit tensile and compressive stresses. On scaffolds with two rows of uprights a spacer should be used to ensure secure anchoring. With tied pole scaffolds the putlogs resting on the masonry can replace the spacer but then the end of the putlog resting on the ledger should be secured against lengthwise displacement. Further, every putlog adjacent to an upright should be secured.

To avoid displacements, scaffolds should be braced longitudinally, and independenet scaffolds should have transverse bracing as well. For bracing, scaffold poles can be used, and for light scaffolds, as those for plasterers, boards will do, but they should be at least 30 mm thick (for the rest see section 2.5.1).

(c)   Ladder scaffolds

Ladder scaffolds are made of scaffold ladders and special parts. The flooring is laid directly on the ladder rungs without intermediate supports so that the spacing of the ladders will depend on the strength of the floor. Connections may be screw hooks or hexagonal screws with extended threads and washers, or ladder clamps, hooks and pins.

According to the rung arrangement scaffold ladders are divided into full-rung, triple-rung, double-rung and single-rung ladders. Scaffold ladders also differ as regards the distance between uprights, that is their width. If the ladder design should be such as to afford the easiest and safest possible passage, then the single-rung ladder should be preferred to all others, for with the rungs 2 m apart unhampered passage is ensured. These ladders, however, are not made to take heavy loads, so the rungs should be reinforced by steel ties whose bent ends are screwed into the uprights; this helps to stiffen the ladder.

With double-rung and triple-rung ladders this reinforcement is not needed. However, with these the passageway is considerably reduced - to a height of 1.5 m with double rungs, and to as little as 1 m with triple rungs. Consequently to move along the scaffold one has to bend under each ladder. With the full-rung ladders there is no passageway at all, hence when they are used flooring should be laid on brackets so that persons can move from ladder to ladder without having to swing dangerously round the outside of the uprights. In any case brackets serve to widen the scaffold floor when the width provided by the distance between the ladder uprights is insufficient for the kind of work being done on the scaffold. With brackets from 25 to 30 cm long for supporting flooring, a scaffold ladder 55 to 65 cm wide can be given an effective width of from 80 to 90 cm. Scaffold ladders of type e less than 85 cm wide are used without brackets, for the available width is sufficient for the light work done on ladder scaffolds, for example plastering or painting.

Scaffold ladders

The wood used for scaffold ladders should be of the best quality. It will generally be tough wood such as red pine, free from knots, not too splintery and not the first year's growth. Scaffold ladders are made in lengths up to 15 m. The rungs mortised into the uprights should be of pine or beech, as far as possible free of knots. At house corners, alcoves, balconies, etc., the ladders should be so erected that they scaffold floor and the edge protection can be continuous inside the ladder uprights. If a scaffold ladder is extended, the upper half should be fastened to the lower with two ladder hooks held by steel ties. In addition the two halves should be fastened together by cross lashing. The overlap should be 2 m.

Anchoring and bracing

Because of its weak longitudinal cohesion, a ladder scaffold needs to be particularly well anchored and braced. Consequently on tied scaffolds every ladder should be anchored to the building at every storey. Anchorages should be so made as not to hamper work and movement on the scaffold floors. The vertical distance between anchors should not be more than 4.5 m. The horizontal distances will depend on the spacing of the ladders, which in no case should be more than 3 m. Ladder scaffolds may be anchored with window jacks and screws or with so-called gable struts and hooks if no window openings are available. When window screws are used special care should be taken if the windows have splayed sides. If the windows have weak mullions (double windows with freestone middle uprights) the anchorage should be carried to the adjacent window opening in order to avoid damaging the mullion by one-sided pressure of the screw. To absorb horizontal forces acting on ladder scaffolds, cross bracing should be installed over every second space between ladders, beginning not more than 4.5 m above the ground and extending to the top.

Flooring

Since the flooring of ladder scaffolds rests directly on the rungs without any other support, the floor boards should be inspected as to their strength and suitability before each use. The boards should have no knots that extend from one side to the other, and should be at least 40 mm thick. For the rest, flooring should satisfy the requirements set out in section 2.5.1.

Edge protection

Edge protection should be provided as required by section 2.5.2.

(d)   Bracket scaffolds on uprights

A wooden scaffold of a special type is the bracket scaffold on poles. It is used only for plastering and painting. The flooring rests on steel brackets, spaced not more than 2 m apart, that are fastened to vertical poles. The ledgers, which are spaced not more than 7 m apart and are fastened to the poles by cords or chains, merely serve as horizontal bracing for the scaffold. Scaffolds over 8 m high should also be cross braced with scaffold poles. The floor resting on the brackets should be at least 60 cm in total width and each board should be at least 25 cm wide.

The steel brackets are fastened to the poles with chains mounted at the top of the brackets, and the bottom of the brackets is shaped like a claw and is driven into the poles. The spacing of the poles will depend on the thickness of the flooring. Scaffold boards less than 3.5 cm thick are not commonly used. Special importance attaches to the method of anchoring the scaffold to the building: the poles should be fastened to firm parts of the building at every second storey.

The same requirements as for pole scaffolds in section 2.5.6(b) apply to the erection of poles, the laying of flooring and the provision of edge protection. Details and measurements are shown in figure 2.154.14.

(e)   Ordinary tubular steel scaffolds

Tubular steel scaffolds are made with steel tubes and special scaffold fastenings called couplings, and usually with wooden floors. These scaffolds can be used for work, for protection or for supporting structures (Fig. 11).

Steel tubes

Steel tubes can be used as uprights, ledgers, putlogs, bracing and guard rails. They usually have an external diameter of 48.3 mm, but the wall thicknesses and the types of steel vary. The specifications of the commonest types of steel tubes are set out in the table below.

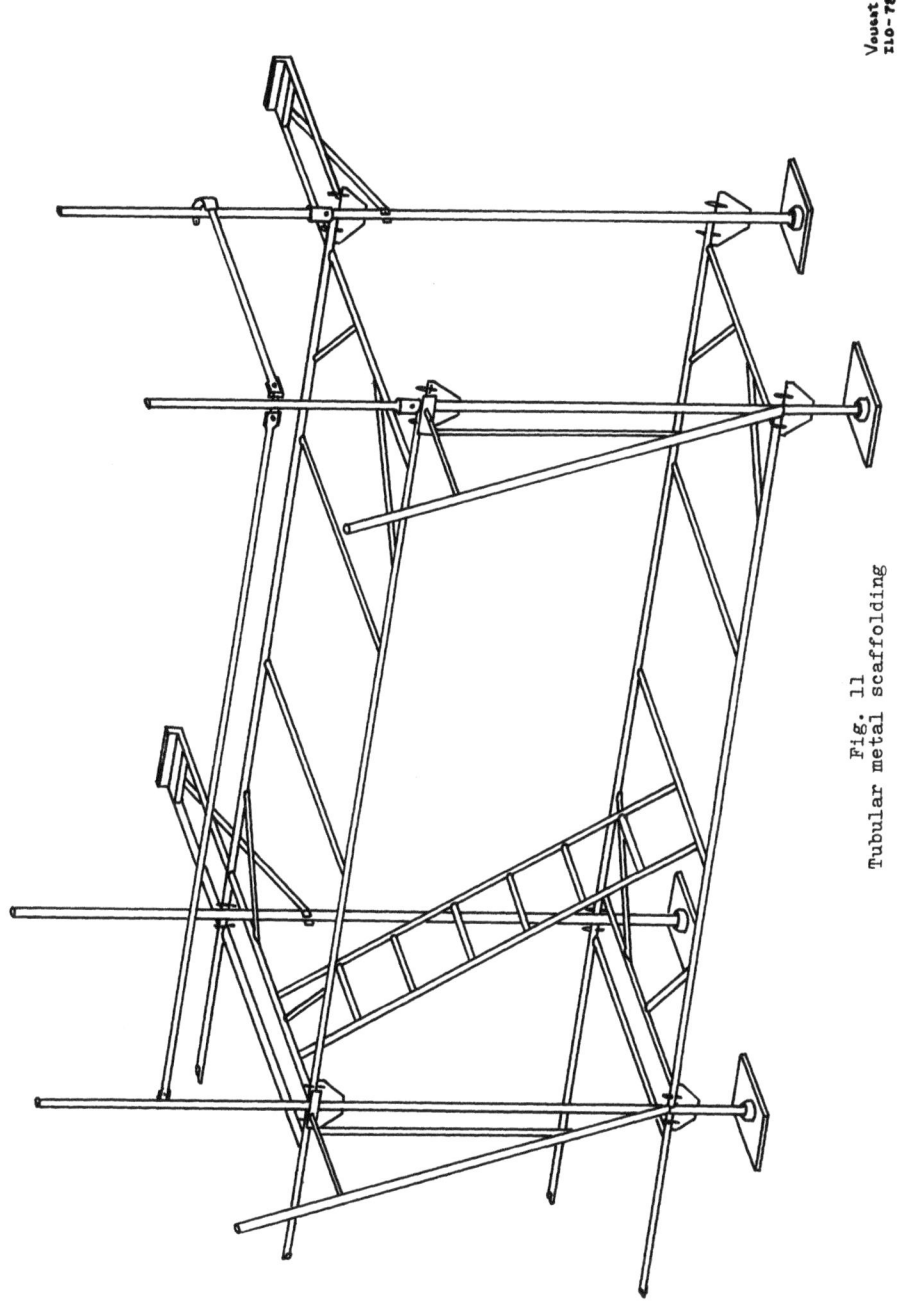

Fig. 11
Tubular metal scaffolding

		Type a	Type b	Type c	Type d
Tensile strength	kg/mm²	35 ÷ 45	55 ÷ 65	35 ÷ 45	35 ÷ 45
Yield point	kg/mm²	21	29	21	21
Breaking elongation A	%	28	17	28	28
External diameter	mm	48.3	48.3	48.3	48.3
Nominal wall thickness	mm	4.00	3.25	3.25	2.90
Tolerance in wall thickness and cross-section	%	+no limit -12.5(1)	+no limit -12.5(1)	+no limit -12.5(1)	+no limit -12.5(1)
Tolerance in external diameter	mm	+0.5	+0.5	+0.5	+0.5
Tolerance in weight	%	+10.	+10	+10	+10

Some countries prohibit the use of steel tubes with a wall thickness less than 3.25 mm because with this thickness deformation due to the force exerted by clamp couplings may be expected.

Tubes should also satisfy the following requirements: the ends should be cut off at right angles to the length; the tubes should be made of corrosion-proof material, or be suitably protected against corrosion both internally and externally. Hot-dip galvanising is better than coating with bitumen, especially for hygienic reasons, because bitumen is very dirty. If tubes are to be welded, the material should be suitable for the welding process to be used.

In order that tubes of different materials and wall thicknesses may be distinguished they should be suitably identified. Only tubes of the same material and the same wall thickness should be used on a scaffold.

Testing of steel tubes

Some countries prescribe tests to ascertain the properties of the material used in tubes, which in general are on the following lines.

The material properties are determined by laboratory tests in recognised testing institutes. In addition bending and crushing tests should be performed.

Bending test. Three tests with three tube lengths. A tube length 80 cm long supported at each end is loaded with 600 kgf at the middle, the load being applied by a rounded stamp (radius 24). In the test the tube should show no lasting deformation, and should sag not more than 3 mm if of type a, and not more than 4 mm if of types b, c or d.

Crushing test. Three tests with three tube lengths. A length of 50 mm is cold-worked with rounded corners in such a way that the distance x the load varies with the wall thickness as follows:

(a) steel with 35 ÷ 45 kgf/mm²: breaking strength: x = 2S;

(b) steel with 35 ÷ 65 kgf/mm²: breaking strength: x = 6S.

On welded tubes, the weld seam shoudl lie at an angle of 45° to the horizontal plane. In these tests the deformed part should have no visible cracks.

Couplings

In addition to various types of tubes, various types of couplings are used on tubular scaffolds. They are mainly clamping devices acting exclusively by friction. So far no satisfactory type of so-called quick-acting coupling on the principle of the bottle stopper or of eccentric coupling has been developed.

Among clamping couplings a distinction can be drawn between:

- screw couplings, whose jaws are clamped on the tubes by the tightening of screws; and

- wedge couplings whose grip results from the insertion of wedges.

Couplings may be classified as follows according to their purpose:

- rectangular couplings: for joining two tubes crossing each other perpendicularly;

- parallel couplings: for joining two parallel tubes;

- terminal couplings: for joining two tubes end to end whether or not the join is proof against tensile stress;

- centring bolts: acting inside two tubes on the same axis to ensure a properly centred join in the erection of the scaffold.

The following general requirements have been laid down for the specifications of couplings. Couplings should be made of suitable material such as steel, malleable cast iron or spheroidal cast iron. The parts should be suitably protected against corrosion, but the inner surfaces of the coupling jaws that come into contact with the tube should not be hot-dip galvanised. Couplings should also be so made that:

- scaffold parts are firmly joined together even under vibration;

- in ordinary use they do not damage the tubes;

- their parts are permanently fastened together;

- the coupling lock does not act directly on the tube;

- it should be possible to adjust locks over a distance of 3 mm when a coupling is fastened to a tube with the minimum permissible diameter in a normal manner.

Testing of couplings

Tests are prescribed for couplings in the same way as for tubes. In general the following principles apply.

Tubes used for testing couplings should be seamless, hot-dip galvanised, and of type a. The pieces needed for testing should not all be cut from the same tube. Screw couplings should be subjected to a tensile moment of 500 cm/kgf; with wedge couplings the wedges should be driven in by three blows from a falling weight that acts with a kinetic energy of 5 kgf/m. The couplings being tested should not be protected against corrosion.

Testing of rectangular couplings

Sliding test. Fifty tests with ten couplings. A coupling fastened to a tube is loaded in its most unfavourable position in order to determine the sliding load PG. After the test neither the coupling nor the tube should show any damage. The sliding load should be 1,100 kgf, and should not be less for more than three tests.

Breaking strength and rotation tests. Five tests with five couplings. A coupling fastened to a massive bar and secured against sliding is loaded in order to determine the displacement . 1 under a load of 1,500 kgf and also the breaking load Pu. The ratio . 1 to the eccentricity e must be $\geq$ 0.123. The average value of Pu must be at least 3,000 kgf.

Clamping test. Five tests with five couplings. A coupling is fastened to a massive bar with a tensile moment of 8 kgf/m. After the test the coupling should show no permanent deformation.

Testing of swivelling couplings

Sliding and clamping tests as for rectangular couplings.

Breaking strength test. Five tests with five couplings. A coupling fastened to a massive bar and secured against sliding is loaded in order to determine the breaking load Pu. The average value of Pu must be at least 1,800 kgf.

Testing of parallel couplings

Sliding tests.

(a) Fifty tests with ten couplings. A coupling fastened to two tubes is loaded in order to determine the sliding load PG. After the test neither the coupling nor the tubes should show any damage. The sliding load must be $\geq$ 1,100 kgf, and should not be less in more than three tests.

(b) Fifty tests with twenty couplings. Two couplings fastened to two tubes are loaded in order to determine the sliding load PG. Test procedure as for test (a).

Clamping test as for rectangular couplings.

Testing of end-to-end couplings

Tensile test. Five tests with five couplings. A coupling fastened to two tubes is loaded in its most unfavourable position in order to determine the load P under which the coupling shifts or breaks. The average of P with end-to-end couplings that are not tension-proof must be $\geq$ 300 kgf, and with tension-proof couplings $\geq$ 5,000 kgf.

Clamping test as for rectangular couplings.

Bending test. Three tests with three couplings. A coupling fastened to two tubes is loaded in its most unfavourable position in order to ascertain its behaviour under a load of 900 kgf. The extension sleeve should show no deformation. The permanent bending of the joined tubes must be $\leq$ 3 mm. The clamping effect should remain after the test.

Sole plates. Sole plates, whether adjustable or not, should be made of suitable material such as steel, malleable cast iron or spheroidal cast iron. The bearing surface should measure at least 175cm². Sole plates should have a device by which uprights can easily be centred. They should undergo a bearing-capacity test consisting of five tests on five plates. A plate laid concentrically on a disc is loaded with 3,000 kgf. It should not break or suffer any permanent deformation.

Structural details of tubular steel scaffolds. As a rule tubular steel scaffolds are erected with two rows of uprights. Scaffolds with a single row of uprights like those with wooden poles should be avoided because satisfactory support for the putlogs cannot be provided in the masonry. Every upright should be placed immovably on a sole plate, and wider distribution of the load should be arranged if the bearing capacity of the ground is low. Joints in uprights should be near junctions with other members of the scaffold, and should be staggered vertically. The ledgers should extend over at least two spaces between uprights and be connected to every upright. Joints between ledgers should not be in a vertical line, but should also be staggered and in addition resistant to tensile and compressive stresses.

All junctions of tubes should be made with couplings. If a number of tubes meet at a point the couplings should be close together so that the tube lengths correspond as far as possible with the mesh of the scaffold, and eccentricities are small.

Tied tubular steel scaffolds should be anchored to the building. The horizontal and vertical distances between anchors should not be more than 6 m, and the anchors should be staggered. For the rest the requirements of section 2.5.4 should be satisfied.

Tubular steel scaffolds should be braced longitudinally, and if they are independent, transversely as well. The bracing at the nodal points of the mesh and at the feet of the uprights should be so fastened that its stresses are transmitted to the ground. Since the stability of a tubular steel scaffold also depends to a large extent on the proper fastening of the couplings, the scaffold erector should be properly instructed in the way to maintain the prescribed initial torque with screw couplings, and in the method of fastening wedge couplings, and should be supervised while he is placing couplings.

(f) **Special types of scaffolds with steel-tube and steel-section frames**

With the object of simplifying and speeding up the erection of scaffolds and so reducing costs, steel scaffolds have been developed in which the uprights have been replaced by vertical frames made of steel tubes or sections, and the ledgers and putlogs, by horizontal frames. As a rule the vertical frames are 2 m high (height of a scaffold level), and the length of horizontal frames corresponds to the usual distance between uprights, which according to the load ranges from 2 to 3 m. The vertical frames are placed one over the other, and the horizontal frames are hung from a suspension mounted on a vertical frame or an upright. The horizontal frames also support the floor. In some systems ordinary scaffold boards are used for floors, in others panels adapted to the length and width of the frames. The railings and bracing consist of tubes, which may also be suspended.

In some systems the construction and width of the scaffold may restrict both the use and the load so that only work requiring little material may be done on it. The use and the height of the scaffold should have been found suitable by appropriate calculations. If some parts of the scaffold are not amenable to calculations they should be tested in a recognised testing institute. The makers have also to certify the soundness of welded joins. All steel parts should be protected against corrosion.

These scaffolds need particularly effective bracing and anchoring because the joints of the vertical and horizontal parts are all in the same plane. Subject to the necessary modifications, the requirements of sections 2.5.2 to 2.5.4 apply to the edge protection, bracing and anchoring. The nominal wall thickness of the steel parts should not be less than 2 mm, or if clamping couplings are used, 3.25 mm. Damaged or deformed parts should be withdrawn from use.

Steel-frame scaffolds should be erected and dismantled under competent direction and supervision. Supervisors must be in possession of the erection instructions of the makers. The maximum permissible height and loading of the scaffold should not be exceeded.

(g) **Trestle scaffolds**

Trestle scaffolds are composed of squared timber or steel trestles supporting a floor. They may be used both for work and for protection. Only properly constructed wooden trestles should be used; they should be sufficiently rigid and the legs should be sufficiently splayed to enable them to withstand vertical and horizontal forces perfectly safely. Makeshift extensions of the legs should not be allowed. Any extensions should be technically sound and proof against compression and bending.

With adjustable trestles the adjustable part should never be pulled quite out of the guides; if should always be rigidly connected to the other part. To block the adjustable part there should be suitably sized pins with a diameter of at least 10 mm fastened to a chain so that they do not get lost. They should not be replaced by unsatisfactory accessories such as nails. If steel trestles are used as adjustable scaffold supports they shoudl be of a stable tripod construction.

Trestle scaffolds should rest on a safe base, and hence not on open joisting. They should not be placed on stones. Not more than two tiers of trestles should be used on a scaffold, and its total height should not exceed 4 m. The trestles should be adequately braced together. With adjustable trestles the adjustable part should also be braced. The spacing of trestles will vary with the thickness of the floor boards.

Type of scaffold	Spacing of trestles with a floor thickness of		
	30 mm	35 mm	50 mm
Maintenance scaffold	1.2	1.5	1.5
Plasterers' scaffold	0.8	1.0	1.2
Masons' scaffold	0.8	1.0	1.0

If the spacing is wider than in the above table, the floor should be supported by beams 100/100 mm or round timber 120 mm in diameter, but then the spacing should not exceed 3 m, or with adjustable trestles, 2 m. Trestle scaffolds higher than 2 m should be equipped with railings and toeboards.

2.5.7 Outrigger scaffolds

Outrigger scaffolds are scaffolds that project from the building and have floors supported by beams, round timber or girders (Fig. 12). The projecting parts may be given additional support by bracing against compression and tension. Outrigger scaffolds may be used for work or for protection. As a rule they are only used as work scaffolds when the imposed loads do not exceed 200 kgf/m². In special cases they serve as supports for other scaffolds, for instance ladder scaffolds and tubular steel scaffolds. When outrigger scaffolds are used as protective scaffolds they should satisfy the requirements for such scaffolds.

(a) Construction

Only wooden beams or steel girders should be used as outriggers. Outriggers should extend into the building at least as far as they project from it. As a rule the length inside the building should be at least 2.5 m. The need to anchor or support them on or against floors considerably reduces their field of use. If the floors have wooden joisting the outrigger should be tied to the joists. At least two lashings with double lashing wire 5 mm in diameter will be necessary. Inside buildings outriggers should be so secured that they cannot lift out, shift or tip over. It is not enough to wedge them in a wall. To concrete floors outriggers should be fastened by two steel hoops 10 mm in diameter embedded in the concrete. The hoops should be anchored by their hooks in the reinforcing rods of the concrete floor. The scaffold should not be loaded until the concrete has sufficiently set. The spacing of outriggers on a scaffold should not exceed 1.5 m. Even at corners of buildings the distance between floor supports should not be greater than 1.5 m. If an outrigger support is in a wall opening there should be a proper supporting construction. If adjacent openings have to be bridged, the intervening masonry pillars should

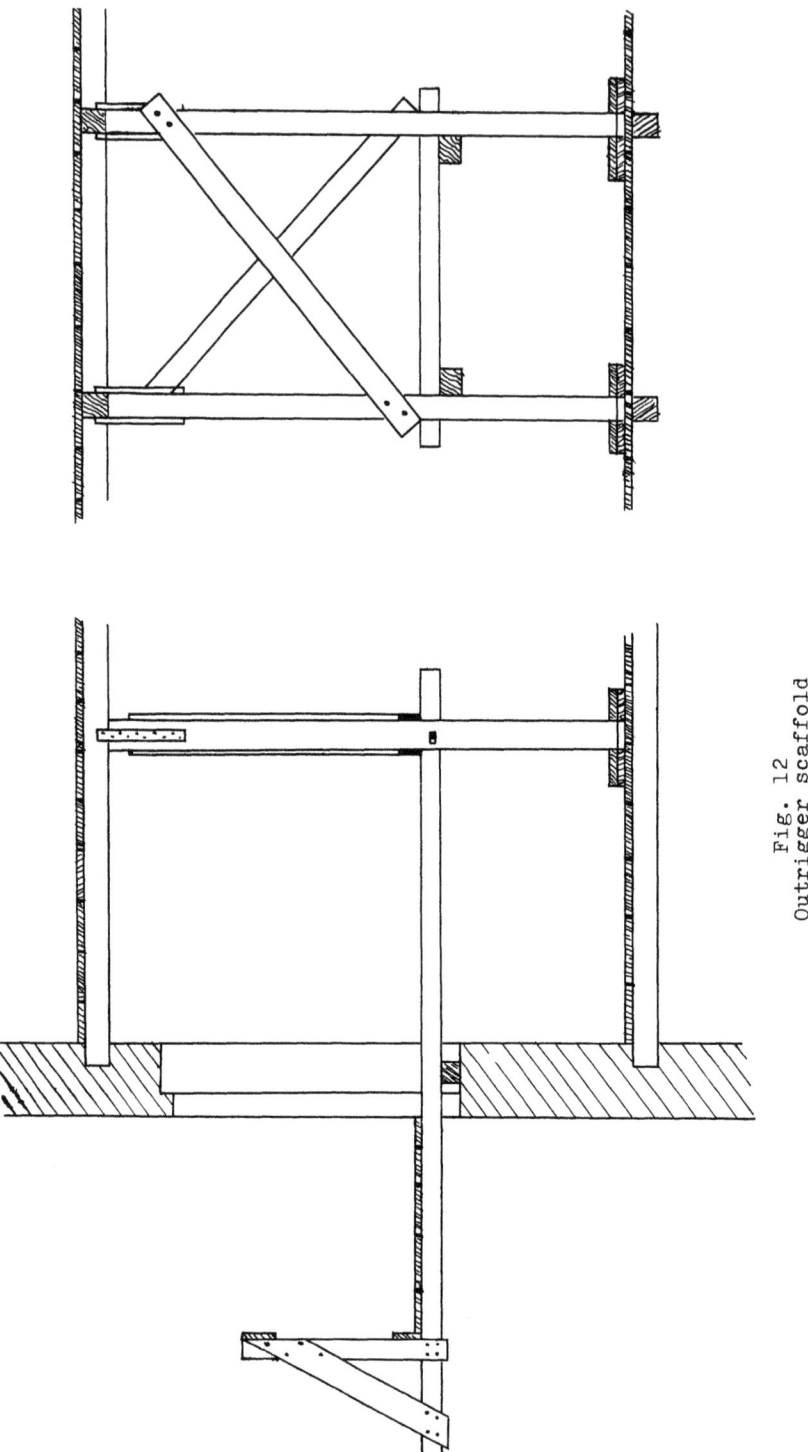

Fig. 12
Outrigger scaffold

be at least 75 cm wide and half a brick thick. For flooring, scaffold boards with a cross-section of at least 250 x 40 mm should be used. Edge protection should satisfy all requirements. The cross-sections of the outriggers can be found in technical tables. Outrigger scaffolds that are used to support other scaffolds should be based on appropriate calculations.

(b) Outrigger scaffolds in conjunction with steel trestle scaffolds

Outrigger scaffolds are sometimes used with steel trestles in such a way that the outrigger rests not on the floor but on an enclosing wall and a steel trestle. The result is a combination of outrigger and trestle scaffolds where the outrigger part serves as a protective and maintenance scaffold, and the part inside the building as a masons's scaffold. As with the simple outrigger scaffold the trestles should be secured to the concrete floor by two steel hoops of 10 mm diameter with their hooks anchored in the reinforcing rods of the floor.

For I NP standard sections and a load of 150 kgf/m² or two loads of 75 kgf the following dimensions are applicable when the outriggers are 1.5 m apart.

Lengths

Scaffold width	Normal outrigger	Corner outrigger 1+2	Corner outrigger
1.00 m	3.80 m 1	4.00 - 4.20 m	5.20 m long
1.30 m	4.10 m 1	4.30 - 4.50 m	5.50 m long
1.80 m	4.50 m 1	4.60 - 4.80 m	5.80 m long

Section thicknesses

	Scaffold width		
	1.00 m	1.30 m	1.80 m
Outrigger I NP	8	10	12
Corner outrigger I NP	8	10	12
Anchor hoop for outrigger or combination trestle with two hoops	⌀ 10	⌀ 10	⌀ 10

2.5.8 Suspended scaffolds

(a) Fixed suspended scaffolds

Fixed suspended scaffolds are scaffolds with floors resting on steel sections, steel tubes, round timber or beams suspended from steel hooks, wire ropes, chains or steel sections. As a rule they are used for maintenance work requiring little material.

The supports from which the scaffolds are suspended should be of ample dimensions. Care must be taken that the supports are suitably positioned on building parts of sufficient bearing capacity.

The following dimensions are applicable with working loads of 100 kgf/m² or two persons standing close together:

	Ledgers	Putlogs
Dimension Diameter	120 mm	160 x 50 mm
Spacing	2 m	1 m
Length supported	3 m	2 m

The floor should be at least 30 mm thick and rest on putlogs, which in turn will rest on ledgers suspended from hooks, chains or wire ropes. Ledgers should be so joined together that they overlap at least 1 m and are tied together by wires, chains or the like. Every join should be separately suspended. Joins should not be made between supports.

If hooks are used as supports care should be taken that the ledgers sit securely in them. The nature of the suspension point may make it necessary to provide a wooden wedge or shim for the upper hook. With open hooks the jaw should be secured against bending. Since hooks are easy to hang up they are generally preferred as means of suspension but tested chains are also suitable. If wire ropes are used the fastening should not be knotted. With wire-rope lashing at least five rope clamps should be used. Wire ropes and chains should not be led over sharp edges.

Since the rupture of a bearing part can lead to a serious accident the greatest care should be taken that they are perfectly sound. Consequently before every occasion of use they should be minutely examined for damage and hair cracks. For safety reasons, for every suspension two mutually independent supports should be used, either of which is alone able to hold the load on the suspension point.

To avoid overloads, such as occur more especially inside buildings where these suspended scaffolds are usually employed, a notice showing the maximum permissible load should be displayed at a conspicuous place. Suspended scaffolds should be provided with edge protection (two-rail railings and toeboards).

(b) Mobile suspended scaffolds

Mobile suspended scaffolds, sometimes called flying scaffolds, have the floor resting on wooden or steel hangers that are hung from outriggers by fibre or wire ropes and can be raised or lowered by blocks and tackle or winches. The dimensions of the individual scaffold parts, including ropes and winches, should be suitable for the use to be made of the scaffold and should be based on appropriate calculations. The outriggers from which the scaffold is to be suspended should be securely anchored in the building and prevented from moving, tipping or overturning. The factor of safety against tipping should be three. Unless the scaffold can be suspended from firm parts of the building, counterweights should be firmly secured to the outriggers. Outriggers should have load-distributing bases on their supports.

If it is more than 3 m long the floor should rest on three hangers; for shorter scaffolds two hangers will be sufficient.

On hand winches, ratchet winches or winches with screw pressure there should be a brake which automatically blocks the load when the handle is released. If mobile suspended scaffolds are used on very tall buildings, winches with special drums should be provided to accommodate the long ropes needed. The speed of raising and lowering will be limited by the manual power needed which will be determined by the transmission ratio but should not exceed 15 kg.

In the working position the scaffold floor should always be horizontal; it should be enclosed on all sides with railings and toeboards.

2.5.9 Suspended scaffolds for roof work

For roof work that is not associated with any other building work, and when ordinary scaffolds are not needed, roofers' suspended scaffolds may prove very useful. They serve tinsmiths as work scaffolds for putting in gutters, and roofers as protective scaffolds. They consist of hangers supporting a floor. Various types of hanger have been developed; they differ in shape and in means of supspension.

As a special type of scaffold the roofers' suspended scaffold should have its bearing capacity certified, and it must also be tested. As a rule the suspension appliances of the different systems are designed for fastening to rafters with the scaffold hanging under the eaves. There may be difficulties in mounting the scaffold if no rafters are available.

One of the most important requirements is that the hangers should only be fastened to sufficiently firm parts of the roof construction. When old roofs are being repaired care should be taken to choose perfectly sound wood for attaching the suspension appliance. With every suspension there should be a second suspension that can by itself take the load to be imposed on the scaffold; a fibre rope, a wire rope or a chain may be used for this purpose. The spacing of hangers will depend on the thickness of the floor boards; with boards 25 cm wide and at least 40 mm thick the hangers should not be more than 2.5 m apart.

Hangers should only be used at places where they do not have to be altered in shape. The bottom wall supports, usually adjustable, should only rest against firm parts of the building. Any makeshift lengthening of wall supports is inadmissible because it would completely change the interplay of forces.

Although the roofers' suspended scaffold can be considered safe if the principles stated above are observed, its erection and dismantling are attended by considerable risks, and consequently workers engaged in these operations should be equipped with a safety belt and life line.

2.5.10 Bracket scaffolds

(a) **Bracket scaffolds for building operations generally**

Bracket scaffolds are scaffolds with the floor resting on brackets hung on the building. They are mainly used as protective scaffolds. The brackets, which are made of steel sections or tubes, should have been based on appropriate calculations. The vertical load is usually taken as 100 kgf/m². All steel parts must be protected against corrosion. Welds should only be made by properly qualified persons.

Brackets, if at heights, should be suspended only from reinforced concrete floors. There should be a double suspension from two wires of at least 8 mm diameter bent to form eyes. The bent wires should be fastened to the reinforcing rods of the floor. The brackets should not be mounted until the floor concrete has properly set. If the wires cannot be fastened to the reinforcing rods the scaffold should not be used. The brackets should not be more than 1.5 m apart. Special corner brackets are needed for bracket scaffolds at corners of buildings.

(b) **Bracket scaffolds for chimney construction**

Bracket scaffolds for chimney construction are hung round the chimney from locked-coil ropes. They are only used for maintenance work. The strength and stability of the brackets should have been based on appropriate calculations. Brackets should not project more than 1 m and not be more than 1 m apart on the outside. They should be equipped with two hooks for suspension from two ropes. Each hook and its fastening should be able to hold the full load imposed on the bracket. The wire ropes to be hung round the chimney for holding the brackets should each be able to hold the full load.

The floor should be made of boards at least 20 cm wide and 30 mm thick, and be secured against falling down, shifting of lifting out. For edge protection taut ropes will do. The posts should be securely fastened to the brackets.

2.5.11 Load-bearing scaffolds

Load-bearing scaffolds are scaffolds with which steel sections, steel tubes, concrete form bearers, beams or round timber directly support loads. They are chiefly used as inside work and protective scaffolds in the construction of silos, chimneys and towers. The dimensions and spacing of the supports will be determined by the method of working, the loads to be imposed and the thickness of the flooring. If the cross-sections and spacing of supports cannot be found in tables they will have to be calculated. The supports should be let far enough into the masonry to ensure a safe bearing surface. The length embedded will vary with the type of support, but it must be at least 6 cm. The bearing surface of the supports (tubes or concrete form bearers) should be of such size that the maximum permissible compressive stress on the supporting masonry is not exceeded.

2.5.12 Shuttering (forms) and its supports

(a) Floor shuttering (forms)

Floor shuttering (forms) has to support the fresh concrete and the reinforcement, in addition to the live load imposed by the dumping of concrete and the moving of workers and equipment. It is carried on bearers resting on supports. Bearers may be beams or more recently extensible appliances of steel; supports may be of round timber or be extensible appliances of steel.

Wooden shuttering supports (forms)

Wooden supports for shuttering (forms) should be straight round timber props in one piece. Their spacing will depend on their bearing capacity, which in turn will depend on the buckling length, and fibre-crushing factor for the wood being used. For the prop lengths usual in floor construction the following maximum permissible loads in kgf/cm² are applicable. When the calculated load exceeds the value presented in the table by more than 10 per cent then the next largest prop should be used. Props over 3.00 m may be used where the slenderness ratio, length/smallest cross-section dimension, has been properly reduced by horizontal cross bracing (lacing) in both directions at intervals not to exceed 2 m vertically.

Un-supported length in m	Diameter in cm (cross-section in cm²)					
	8 (50 cm²)	10 (80 cm²)	12 (115 cm²)	14 (155 cm²)	16 (200 cm²)	18 (255 cm²)
2.50	0.85	2.22	4.30	6.88	10.00	13.60
2.60	0.78	2.03	4.04	6.64	9.72	13.25
2.70	0.71	1.87	3.85	6.38	9.40	12.92
2.80	0.66	1.72	3.65	6.12	9.14	12.70
2.90	0.60	1.58	3.39	5.84	8.80	12.40
3.00	0.56	1.46	3.20	5.58	8.55	11.92

As already mentioned in previous sections, in additon to the weight of the floor and the shuttering, account has to be taken of other loads when calculating the total load to be imposed. These loads include wind loads and live loads. As the figures in the table show, the maximum permissible load is inversely proportionate to the length of the props, and directly proportionate to their cross-section. Props less than 8 cm in diameter should not be used. The regulations of some countries require a larger minimum diameter for wooden props. Some countries restrict the minimum prop size to 4" x 4" (10 cm x 10 cm) for sawn timber props.

Under caps only every second, and under beams only every third, prop should have joints. The joined props should be evenly distributed. Props with more than one joint should not be used. The end surfaces of joined props should be horizontal and fit well together. The joints should be secured against buckling and overturning by nailing on ties at least 70 cm long; for round timber props there should be three ties, and for square props, four. Joins should not lie in the middle third of the props.

Extensible steel props

These props usually consist of a pipe on a sole plate and an extension with a cap or some other piece for supporting stringers. The extension arrangement allows for coarse and fine adjustment. The bearing capacities for the various extended lengths should have been properly calculated. The means of extension may require determination of the bearing capacity of props to be based on tests. These tests will serve to determine the following critical loads:

- the maximum load attainable;

- the load that causes elastic buckling of 1/500;

- the load that causes either the prop to be permanently shortened by 2 mm, or to acquire a permanent buckling of 1/500.

In the tests, following the common practice, it is to be assumed that the props are secured only at the top and bottom against lateral displacement. Since in practice a centrally applied load cannot be counted on, in the test an eccentricity should be assumed whose magnitude is found from double the core diameter of the extension piece in the full cross-section, the value 1/500 and the extended length in question. This magnitude of the assumed eccentricity corresponds to the forces affecting the loading of the prop axis and due to the condition and length of the prop and any inadvertent tilting.

The following table gives a figure for the bearing capacity of an extensible prop with a yield point at least double the strength, when the outer pipe has an external diameter of 60 mm and a wall thickness of 3.5 mm, and the extension pipe an external diameter of 48.3 mm and a wall thickness of 3.25 mm.

Extended height in m	Maximum permissible load in t
2.4	3.1
2.7	2.5
3.0	1.8
3.3	1.3
3.6	0.9
3.9	0.6
4.2	0.35

Here also the bearing capacity of props decreases greatly as their length increases.

As regards their construction extensible props should satisfy the following requirements:

- because of the possible denting of metal props in the rough conditions on building sites, the minimum wall thickness of the load-bearing cross-section of the prop should be 2.5 mm;
- the steel used should be guaranteed to have a yield point at not less than 2,400 kgf/cm^2;
- the minimum thickness of the cap and the sole plate should be 8 mm;
- the guide length of the two halves of the prop should not exceed 30 cm at the maximum mechanically attainable extension, and the halves must be mechanically secured against pulling out so that it will not be possible to insert extension pieces that are too long for the type of prop in question.

Placing and bracing of props

Loads on props should be properly distributed at ground level. Special measures should be taken if the ground is soft or frozen. To ensure adequate distribution of the load on the ground the props should have a secure and immovable base such as a beam or a board. They should not be placed on loose bricks, barrels, cases or the like. If more than one layer of blocks, etc., cannot be avoided they should be secured against tipping.

Props should only be placed with the entire sole plate resting on a flat surface. If they are to be inclined they should rest on sufficiently large and suitably sloping wedges.

The stability of the structure formed by the props should be ensured by adequate bracing and lacing.

In order to enable the shuttering to be removed safely and without jolting or vibration, wooden props should be placed on wedges. When steel extensible props are used, smooth withdrawal can be achieved by altering the fine adjustment.

When the floor area to be concreted is not bounded on all sides by firm walls or pillars, or when it is very large and unsupported, the props should be braced longitudinally and transversely. Such bracing is also indispensable when high floor shuttering is supported on props.

Shuttering (form) bearers

Instead of stringers consisting of beams supported on a number of wooden props, shuttering bearers of adjustable length have been developed that only need to be supported at the ends, or if the span is long, in the middle as well. To determine the bearing capacity of these adjustable bearers, suitable tests should be undertaken to find the tolerable bending moment, the resistance to bending in the most unfavourable combination of bearer and extended length, and also the tolerable transverse stress and stress on supports. The results of the tests together with appropriate calculations can be used to compile loading tables. In these tables account is usually taken of an additional load, that of the shuttering itself and its bearers.

The following requirements apply to the construction of shuttering bearers:

- for steel and light metal bearers the minimum wall thickness of cross-sections is 2 mm;
- the bearers should be protected against corrosion;
- bearers should only be placed with their claws on masonry when its bearing capacity has been established. Otherwise they should be placed on round timber uprights or on other load-bearing parts of the shuttering supports. If intermediate supports are necessary it must be ensured that the bottom girder of the bearer can withstand the additional stresses that their use would cause, and if they cannot the girder must be suitably reinforced.

(b) Supporting scaffolds for arches, etc.

A distinction should be drawn between wooden and steel supporting scaffolds.

Wooden supporting scaffolds should be properly carpentered and engineered, assembled and installed. If unseasoned wood is used, or wooden parts are permanently under water, the maximum permissible stresses should be reduced in accordance with official regulations. The slenderness of wooden tower scaffolds is expressed as the ratio of tower height to support spacing and should not exceed 8. For considerable heights intermediate ties should be installed. For tower connections expanding plugs or simple packing should be used as far as possible and their bolts should be fitted with washers. Bolts should be tightened up from time to time and especially shortly before applying the load.

On multi-level scaffolds the props should be so arranged that the load on upper props will be transmitted directly to those underneath.

When steel is used, the structural elements should be so marked that elements of different bearing capacities cannot be confused. Bent and otherwise damaged couplings, spindles and the like should not be used. Erection of the scaffold should be governed by the rules for structural steel erection.

As regards flooring and edge protection, the requirements of sections 2.5.1 and 2.5.2 apply.

For the erection, use and dismantling of supporting scaffolds, see section 2.5.

On tightly assembled steel supporting scaffolds deformations under differing load conditions may give rise to additional stresses, as a result for instance of eccentricities of the forces acting on supports and connections, unexpected horizontal stresses on buckling lengths, and crushing forces on certain members. These matters need careful attention, because they may lead to overloading the scaffold.

The slenderness of steel tower scaffolds expressed as the ratio of tower height to support spacing should be ≤ 10. For considerable heights intermediate ties should be used. If so-called scaffold appliances made of steel are used (multi-unit supprots and bearers) it should be established for every kind of use that the maximum permissible load as determined by calculations and tests is not exceeded.

With multi-unit props the loads should be centred by suitable shaped pieces at the top and bottom of the props so that the load is evenly distributed over all the units. If shuttering bearers are used strength calculations should always be made for each type. The safety factor against breaking should be at least two, and for couplings and sole plates, at least three.

The necessary diagonal bracing should be based on strength calculations. Tubes should only be extended by methods that ensure adequate resistance to tensile and compressive stresses. Tube joints should be as close as possible to the nodal points of the mesh.

For the lowering devices requisite on supporting scaffolds, a safety factor of three against breaking should have been certified. Under full load spindles should not project more than four times their nominal diameter. With long spindles the horizontal forces should be deflected by suitable lateral supports or assemblages.

Erection, use and dismantling of scaffolds

2.5.13 General

Only good and suitable materials and tools should be used for the erection of scaffolds and they should be examined as to their fitness for use before every occasion of use. Care should be taken that the qualities of wooden and steel parts conform to the relevant requirements.

Scaffolds should be erected and dismantled only under competent direction and by competent workers. During erection and dismantling all other work in the immediate vicinity of the scaffold should stop unless special precautions are taken. The persons engaged in these operations should be familiar with them and competent to perform them. All workers should be provided with safety belts and should be required to wear them. During erection and dismantling only persons engaged in the operation should have access to scaffolds. Scaffolds should not be used before they are complete. To ensure this, even scaffolds that are only temporarily incomplete should be indicated by a conspicuous notice.

Scaffolds should be under continuous supervision, should be inspected after lengthy interruptions in the work and after storms; connections and anchorages in particular should be tested. Work on suspended scaffolds should stop during storms. Every scaffold floor in use should have safe means of access, usually inclined ladderways or inclined gangways. Scaffolds should not be overloaded. Building and demolition materials should not be stored on scaffolds in excessive quantities. On work scaffolds storing of materials is only permissible to the extent necessary for the work in progress and allowed for in the construction of the scaffold. Persons should not jump on to scaffold floors. In icy and frosty conditions scaffold floors and means of access should be strewn with sand or the like.

2.5.14 **Additional requirements for shuttering and supporting scaffolds**

Additional requirements should be imposed on the erection and dismantling of supporting scaffolds. It should be possible to remove shuttering easily, safely and without jolting and vibration. It should therefore be mounted on wedges, sand buckets, spindles or other supports. Before the concrete is poured into the shuttering all parts of it should be examined once more as to their stability. Also while the concrete is being poured the scaffolding should be constantly watched to see whether the props and their connections have shifted. No shuttering or support should be removed before the concrete or the masonry has set sufficiently. The responsible builder should satisfy himself of this before he orders any dismantling.

The periods for which shuttering should be left in place are specified as follows in the regulations for reinforced concrete construction.

Type of cement	Side shuttering of beams and shuttering of walls, columns or pillars	Shuttering of floor panels	Supports of beams and large floor panels
Cement 275 and trass cement	3 days	8 days	3 weeks
Cement 375	2 days	4 days	8 days
Cement 475	1 day	3 days	6 days

As a rule columns and pillars should be stripped before beams and panels. Usually props, centring and wide-span floor shuttering should be lowered by cautiously releasing the fastening appliances. They should not be pulled off violently and all vibration should be avoided.

While shuttering is being removed no one should remain unnecessarily on the shuttering or the scaffold. Access to the area should be barred to persons not engaged in the work.

2.6 Working platforms

General

The expression "working platform" covers equipment of the most widely differing kinds; the only thing that they have in common is their use as a workplace, usually for work of short duration done at a certain height for which the erection of scaffolding would be uneconomic and time consuming. According to their type, working platforms may be compared with scaffolds or lifting appliances. Platforms that are fixed, move on wheels (Fig. 13) or hang from jibs are comparable with scaffolds. They only differ from scaffolds in that they cannot be erected in different ways with varying numbers of units like building blocks according to the particular use to be made of them but can only have one surface area and usually only one height for all purposes. But working platforms that move over walls such as those of facades, silos and chimneys and are usually designed to carry one person resemble lifting appliances for passangers, and some are built and operate like protective devices for elevated workplaces. This variety of types makes it impossible to go into every detail in the following paragraphs, which are therefore confined to general recommendations.

By reason of their uses all these appliances should satisfy strict requirements as to their strength, stability and general safety. Accidents at working platforms are mainly due to misuse, and attributable to the fact that the users were not sufficiently acquainted with their construction and method of operation and not sufficiently instructed in the rules to be observed. Defective and unworkmanlike anchoring of the appliances is another cause of accidents. Constructional defects are less often the cause of accidents because nearly all platforms have to undergo a type test or require a type authorisation.

Fixed working platforms and platforms moving on wheels

These platforms are usually intended to carry one or two workmen with their tools and materials. A distinction can be made between fixed platforms standing on uprights, and platforms that move up and down on telescopic uprights. They are usually built to take loads of 100 kgf/m², and if some carrying parts have to bear heavier loads, these are calculated as two loads of 75 kgf 500 mm apart. On the assumption that the stability of the platform when used in the open, or on openwork constructions, or when exposed to wind pressure, or to exceptionally large lateral forces (for example when pneumatic or similar tools are used) is ensured by anchoring proof against tensile and compressive stresses to the building or parts of it or equivalent supports or by taking equivalent precautions, a stability safety factor of 1.5 is sufficient for a horizontal force of 30 kgf applied at the height of the platform in the most unfavourable conditions of loading without wind pressure. Other precautions include workman-like guying with ropes, and jibs that can be swung out and secured with spindles to enable the platform base to be broadened. In any case the platform must be adequately secured against tipping over.

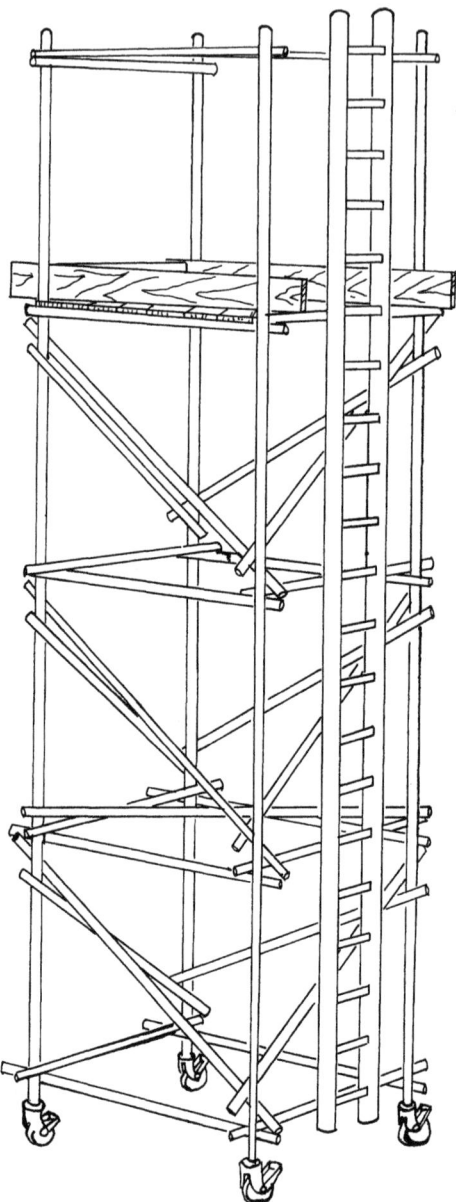

Fig. 13
Mobile tubular metal work platform

The floor and edge protection of a working platform should be the same as for work scaffolds. Access by ladders should be possible: climbing up the building should not be allowed. Travelling platforms should be moved only when they are empty and they should be secured against running away. See also under travelling pole scaffolds, 2.5.5(c).

Working platforms travelling vertically

These include hand and power-operated platforms designed for one or two, and less frequently three, persons, and fastened by ropes to beams projecting from buildings. The winches should be safety winches that can automatically hold the load in any position.

Instead of winches so-called grippers may be used; they should be equipped with a locking device. The ropes should have a safety factor of at least eight. The strength of all parts of the working platform, the suspension, and its anchorage on the building should have been tested and certified.

Working platforms that travel vertically should only be erected under competent supervision. While erecting the platform workers should use safety belts. Before every occasion of use, all parts of the platform, including the winch and ropes, should be examined for good working order and safety. Platforms should only be raised and lowered when it has been ensured that they can move freely. While work is being done a platform should be secured to the building. During storms and squalls work should be stopped and the platform lowered. Ladders should not be used on platforms.

Raising and lowering appliances

A special group is formed by work lifts which, instead of a platform, have a seat or a basket for carrying persons. They are raised and lowered on a rope by hand-powered equipment. These appliances are only used for small jobs that do not require much in the way of materials and tools. On these appliances the manufacturer should clearly mark the maximum permissible load and the dead weight. The suitability and quality of the material should have been certified. The lifting mechanism should satisfy the requirements for winches (section 1.3.1(f)). The ropes should have a safety factor of at least eight, be resistant to corrosion, and not twist much. The end attachments of the ropes should not be ordinary clamps, but talurite clamps with thimbles. All fástenings, including those for the suspension, should be secured against inadvertent loosening, for example by a pin and a lock washer on nuts.

Seats and baskets should be so made that the user and his tools and materials are effectively prevented from falling out.

The appliance should be assembled and used in strict conformity with the manufacturer's instructions, which should be available at every place of use. The bearing capacity of the suspension and its anchorage on the building should be certified in every separate case and the certificate should cover the bearing capacity of the structural part from which the appliance is to be suspended. The bearing capacity should be calculated with a safety factor of three. Raising and lowering appliances should only be erected, dismantled or altered by experienced persons under competent supervision. These persons should wear a safety belt. The maximum permissible load should not be exceeded, and inclined lifting should not be allowed. The user should be familiar with the appliance, its mechanisms and its method of operation. Before any raising or lowering of the appliance it should be ensured that it can move freely. During storms and squalls work should be stopped. If visible defects are noticed in use the appliance should be immediately lowered and laid off. Appliances should be maintained in accordance with the manufacturer's instructions.

2.7 Ladders, ladderways, stairs, gangways

Ladders

2.7.1 General

Falls from ladders rank high among the common causes of accident and particularly in the building industry the injuries are often serious. Many accidents are caused by defective ladders but more by mistakes of the users. If fixed stairs were installed whenever conditions allowed, ladders would not be needed and the root cause of the accidents would be eliminated. However in many building operations the use of ladders is unavoidable and it is all the more important that they should be provided and maintained in perfect condition; at the same time every effort should be made to train all users in the safe use of ladders.

2.7.2 Types of ladders

(a) Technical requirements and definitions

Ladders may be made of wood, steel or light metal and latterly synthetic materials have been used. They must conform to recognised technical rules as regards materials and manufacture. Standards, recommendations, directives or regulations have been issued in some countries to ensure that manufacturers meet these requirements.

All kinds of ladders are used at building operations. They may be classified by type and purpose:

Portable single ladders: ladders that can be used only when leaning against some structure or object.

Step ladders: independent double ladders hinged at the top.

Sectional ladders: ladders consisting of a number of sections assembled together.

__Extensible ladders__: ladders that can be extended by pulling out one or more parts.

__Multi-purpose ladders__: ladders in which two or more types are combined.

__Fixed ladders__: ladders fixed in position and usually vertical.

__Suspended ladders__: portable ladders fitted with hooks that are used only when suspended.

__Mechanical ladders__: mobile extendible ladders that are extended by winches.

(b) __Portable single ladders__

General

On wooden ladders the rungs should not be merely nailed or screwed in. They should also be recessed and the uprights should be strong. However in some countries this method of fastening rungs is not allowed, and in these, in addition to being nailed or screwed into the uprights, the rungs are secured by lengths of wood running along the uprights between the rungs.

At building operations ladders of all lengths will be needed to reach workplaces at many different heights. When deciding how long a portable ladder is needed it must be remembered that it should extend 1 m above the top landing unless another handhold is provided for the user as he steps off. If to obtain the necessary length two ladders are tied together the overlap should amount to one-fifth of their combined length, and the upper ladder should lie under the lower. The spacing of rungs should be the same on both ladders for otherwise they cannot be used safely.

Portable single ladders should be secured against excessive sagging either by design or by reinforcement. They should only be placed against firm supports. Panes of glass, wires, posts and open doors are not safe supports.

Placing the ladder

The safety of a person using a portable ladder depends largely on the angle at which it is set up. Practical tests and technical calculations have shown that the ladder is safest and easiest to climb at about 75° to the horizontal. This angle is obtained when the horizontal distance from the top support to the foot of the ladder is one quarter of its length. The correct inclination can also be found as shown in figure 2.171.222.

Securing ladders against slipping and tilting

It is particularly dangerous if a portable ladder slips or tilts; to prevent these hazards ladders should either be securely fastened or provided with non-slip feet (Fig. 14). If neither of

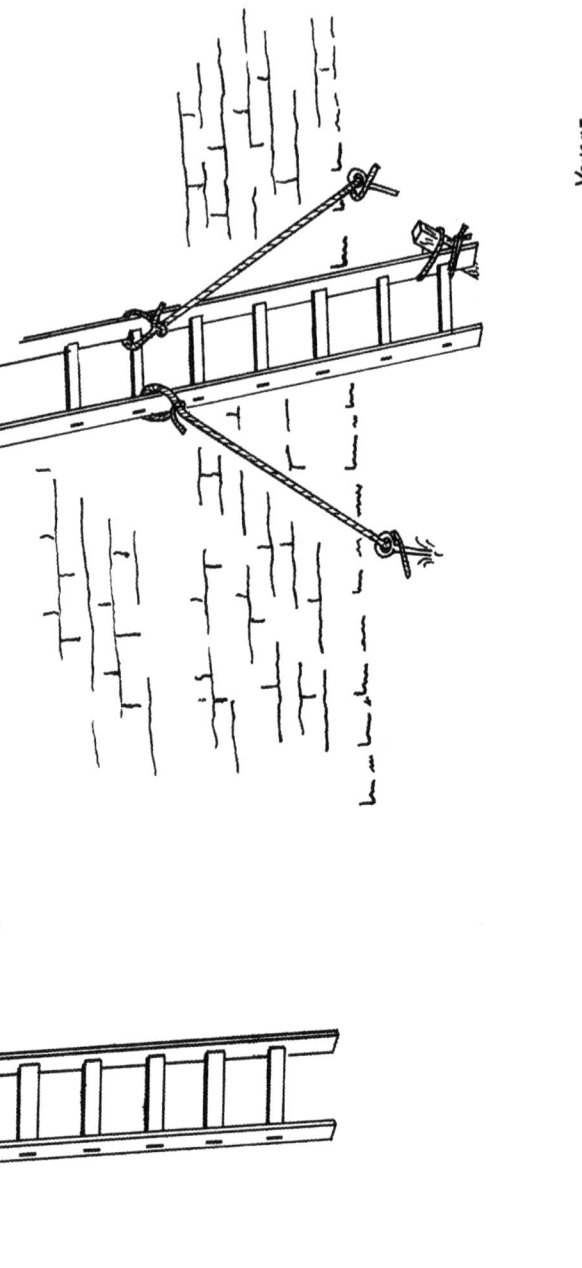

these precautions is practicable the ladder should be held by a second person, but this is not a reliable protection as the man at the foot cannot stop the ladder once is starts slipping. The best protection for portable ladders is to fasten them as this will prevent them from slipping or tilting. A ladder only fastened at the bottom is prevented from slipping forwards but not sideways. Since ropes for fastening are often not to be found when they are wanted, it is advisable to tie one to the ladder upright so that it is always ready for use. The risk of slipping can also be met by providing ladder shoes but this does not give complete protection, because their efficacy depends on the nature of the ground. The risk of slipping on smooth ground can be reduced by placing a rough board under the ladder. Among other precautions are fixing a plank in front of the ladder feet, and clamping the ladder to the ground.

(c)   Step ladders

In addition to portable single ladders, step ladders are used at building operations when independent ladders have to be used inside buildings. These step or trestle ladders can be climbed from two sides. The two sides should be secured against coming apart. However limitation of spread should not be ensured by a stop on the top of the ladder because when the ladder is opened the two sides rest against each other and counter bearings are formed over the hinge. This type of construction is prohibited in some countries because in the course of time the hinges work loose, and this leads to splitting of the sides by the screw holes.

Coming apart can be prevented by chains or toggles, which must be permanently attached to the ladders half-way up. Hemp ties should only be used on ladders that are not exposed to damp. String may not be used instead of ties or chains.

(d)   Fixed ladders

Fixed ladders are usually fixed permanently, but they may also be used for climbing a scaffold that has only one space between uprights so that a single inclined ladder cannot be laid alongside it. Fixed ladders for climbing cranes, silos, buildings, etc., should be securely fastened in position. The fastening points should be spaced not more than 2 m apart, and the top and bottom points should not be more than 75 cm from the end of the ladder. The uprights should be at least 25 cm apart. The clearance between the rungs and any fixed structural part behind them should be at least 15 cm. As with portable single ladders, the uprights of fixed ladders should extend at least 1 m above the top landing, unless another handhold offers sufficient protection against falling.

Climbing vertical ladders, especially if they are very long, is awkward and tiring, and so possibly unsafe; ladders over a certain length should be equipped with a back rest beginning 3 m from the bottom. If a back rest cannot be provided because of constructional difficulties or operating conditions as at masts a holding device can be used that in ordinary climbing up and down runs along a pipe bearing catches. The person on the ladder wears a belt to which the holding device is fastened. If he falls the device is seized by the next catch and so the fall is arrested.

(e) **Extension ladders**

Extension ladders are merely a special form of portable single ladder and are used like them, so that they should satisfy the same safety requirements. Consequently in every position of use they should possess the same solidity as ladders of the same length with continuous uprights. The spacing of rungs should be the same in all sections of the ladder. The overlap of one-fifth of the ladder length required for joining two ladders is not necessary with extension ladders. The sleeves should be so shaped that the extension is subjected to some stress. As a rule the rung surfaces are so close together that there is no risk to the user when the upper part rests on the lower part and there is no need to comply with the requirement that the upper part should be under the lower part (which applies when two ladders are joined).

(f) **Multi-purpose ladders**

Collapsible portable single ladders have been developed as multi-purpose ladders so that when they are used for erection purposes they can be transported more conveniently; they can also be used as trestle ladders. However, the fastening chain can be attached and detached, so that the requirement concerning permanent fixation is not met. Multi-purpose ladders should be designed to satisfy the requirements for the single ladders they replace.

(g) **Mechanical ladders**

Mechanical ladders are used with great economic advantage, more especially in maintenance work on buildings. The requirements to be satisfied by the design of the ladders are the same as for all types of ladder, namely security against excessive sag, uniform solidity of joined ladders, and uniform rung spacing. The causes of accidents with mechanical ladders lie less in the design of the parts than in defective stability resulting from improper positioning, failure of the winch, or improper use.

Travelling ladders should be provided with devices to secure them against accidental displacement. They should not be moved while anyone is on them. The winch design should comply with the technical rules for winches generally. Mechanical ladders should be equipped with blocking spindles, which have to be let out and fixed to relieve the weight on the carriage before the ladder is extended.

There should be rules for the use of mechanical ladders, covering among other things the limits of use, for instance the maximum permissible extension for a given inclination. The operator and a relief should be trained in the use, inspection and care of the ladder. Before any occasion of use the ladder should be examined to see whether it is in proper working order. The operator should see that users are wearing safety belts unless they are working on a fenced platform. If mechanical ladders have to be repaired the work should be done only by reliable persons who are familiar with such ladders. Tests of the ladders should be made every two years by a competent person and the results entered in a register.

2.7.3 **Other general precautions in the use of ladders**

(a) **Transport of ladders**

As a rule ladders should be carried so that their front end is always above head level, that is 2 m above the ground, so that anyone coming round a corner is not injured by them. Extendible ladders should always be transported unextended. Ladders should always be carefully laid down, not thrown down. If ladders are transported on motor trucks they should rest on soft supports. Materials should never be stored on ladders.

(b) **Climbing ladders**

When climbing up and down ladders a man should always turn his face and never his back towards the ladder; this also applies to working on a ladder. A man should not grasp the sides, but the rungs, with his hands, for if his feet slip off the rungs he will not be able to grip the thick sides well and will be in danger of falling.

A man climbing a ladder should have both hands free for holding on. Any objects needed can be safely carried in a tool bag fastened to a belt or a bag slung over the shoulder. The use of carriers also reduces the likelihood that tools will be left lying about in places where they could fall as a result of vibration and injure people.

Suitable footwear is essential for safety in using ladders; only boots or shoes that give a good foothold should be worn and damaged footwear should never be worn.

(c) **Work on ladders**

A man who works standing on a ladder soon tires in his uncomfortable position and this impairs his attention and reaction capacity. Since work on ladders always involves a certain risk, it should be of short duration. Brackets hung on the rungs will give a good foothold and so make the work easier.

A ladder should never be climbed to the top rung because there is no safe stand there. Similarly a man on a ladder should not lean over sideways too far because the resulting displacement of the ladder's centre of gravity may cause it to tip over.

(d) **Misuse of ladders**

Ladders should not be used for purposes for which they are not suited or are not intended. Many accidents have been caused by using trestle ladders as portable single ladders. Above all ladders should not be used in a horizontal position as gangways, or as platforms or for carrying loads, because they are not designed to withstand the stresses resulting from such use.

(e) Control and maintenance of ladders

The safety of a ladder user depends largely on how well the ladder is maintained. Sooner or later the use of damaged ladders leads to accidents. Consequently ladders should be frequently examined, superficially before every occasion of use and thoroughly at least once a year. Damaged ladders should be taken out of use at once. If a ladder becomes unusable it is best to destroy it at once, but it should be replaced without delay or else unsuitable means will be used for climbing. Damaged ladders that can be repaired should not be patched up but repaired properly by a competent person.

The condition of a ladder depends not least on the way it is stored. Ladders are most suitably stored in sheds where they should rest horizontally on hooks in a wall. Parts of hooks can be shaped as supports.

Ladderways and stairs

Ladderways as ordinarily used on scaffolds for access to the different levels, or in staircases if the stairs have not been installed, should not have the ladders vertically one above another so that objects falling could strike the ladders below. If conditions on the site prevent proper arrangement of ladders, they should be enclosed at the sides and underneath, but not so as to impede movement up and down. Similarly ladderways should be protected if they are erected over a passageway or a workplace.

If ladderways and stairs are used for the transport of building materials they should not be carried more than two storeys in a single flight, and sufficiently wide passing places should be provided.

Like scaffolds, stairs should be protected at the sides with standard railings and toeboards. The parts of the building supporting temporary stairs should be rigidly connected to the landings so as to prevent the stairs from being displaced or slipping. Temporary stairs, and also concrete stairways until the steps have been made, should have treads laid over their entire width at intervals not exceeding 50 cm. Owing to the rough conditions on building sites the treads are easily dirtied and so rendered unsafe. Provision should be made for regular cleaning. Recently, instead of treads temporary steel stairs have been used; it is fairly easy to place them on the stairway.

Gangways

Gangways should be so fastened and supported that they do not slip off their supports, tip over or rock when they are walked on or ridden on. Inclined gangways may be carried to a height of 7 m without a level landing if the slope does not exceed 1 in 5. The slope should never exceed 35° (approximately a slope of 1 in 1.5).

The width of a gangway will depend on the use made of it. If it is only a passageway it should be at least 80 cm wide; if it is used for the transport of materials it should be at least 1.25 m wide; if it is steeper than 1 in 5 it should be provided over its entire width with treads 50 cm apart, as for stairs.

2.8 **Tools**

General

A distinction should be drawn between hand tools and power-driven portable tools. The latter in turn can be divided into electric tools and pneumatic tools according to the source of power and a recent addition to the tool family is the powder-actuated tool or bolt gun. This mere recital of the classes of tools shows how important they are from the accident-prevention point of view and indeed tools account for a sizeable fraction of all occupational accidents. The causes may lie in defects in the tools or in carelessness in the users, but sometimes also in the nature of the materials worked. With power-driven tools, to the dangers of the tools themselves must be added those of the source of power - compressed air or electricity. The chief danger with powder-actuated tools lies in the design of the tool and in its resemblance to a fire arm.

Hand tools

2.8.1 General requirements

Hand tools should be made of good quality material and be suitable for the work; they should only be used for the special purposes for which they were designed.

Wooden handles should be made of best quality straight-grained wood, be of the right shape and size, and have a smooth surface free from splinters and sharp edges.

Percussive tools should be made of carefully selected steel that is hard enough to stand up to blows without mushrooming but not so hard that it splinters or breaks. The heads of percussive tools should have the edges trimmed or ground as soon as they show mushrooming or cracks. Hand tools should only be tempered, trimmed and repaired by competent persons. When not in use sharp-edged or pointed hand tools should be suitably protected.

Hand tools should not be left lying around in workplaces or passageways, or at elevated places from which they could fall on persons underneath. Suitable and readily available containers, holders, etc., should be provided for storing tools. All tools should be examined at regular intervals by a competent person and damaged tools should be replaced or repaired. Every worker should be instructed and trained in the use of his hand tools.

2.8.2 <u>Special requirements</u>

(a) <u>Axes, hatchets, etc.</u>

Axes, hatchets and similar hand tools for working non-metallic materials should always be kept sharp. Their handles should be carefully fitted in the heads and securely fastened to them. Holders of suitable material should be provided if necessary for carrying these sharp-edged tools.

(b) <u>Chisels, centre punches</u>

The blades of chisels and the points of punches should be of a suitable shape for the work to be done and kept sharp. When chisels are used for knocking off rivet heads or similar jobs protective screens or mesh should be provided and used, so that other persons cannot be injured by flying splinters. If the work is likely to endanger the eyes of the user he should wear safety goggles.

(c) <u>Crowbars</u>

The edges and points of crowbars should be kept in good condition to reduce the risk of slipping to a minimum. While crowbars are not in use they should be laid flat and not leant against a smooth surface.

(d) <u>Screwdrivers</u>

The edge must be kept properly sharp, fit well in the slot of the screw head and be free from oil and grease. If the edge is worn down or the handle split, the screwdriver should be withdrawn from use until it has been properly repaired. Screwdrivers should not be used as caulking tools or chisels.

(e) <u>Spanners and wrenches</u>

Spanners or wrenches used for tightening up or loosening nuts should be of a suitable size; attempts should never be made to adjust the size of a spanner by inserting a piece of wood, a nail or other object. Spanners and wrenches should not be used with pipes or other extension pieces unless they are specially made to take them. Spanners with worn jaws or other defects should be taken out of use.

<u>Pneumatic tools</u>

2.8.3 <u>General requirements</u>

Pneumatic tools should be issued only from the tool store or by a competent person and after use should be returned to the issuer. On these occasions they should be given an external examination. They should also be thoroughly overhauled, cleaned and tested at regular intervals depending on the frequency of use. If defects are found they should be kept out of use until they have been properly repaired.

2.8.4 Special requirements

The compressed-air trigger on pneumatic tools should be so designed that the risk of inadvertent starting up is reduced, and so arranged that the compressed-air inlet valve closes automatically when the pressure of the operator's hand is removed.

The compressed-air hose and its connections should be designed for the work to be done and the pressure to be used, securely joined to the pressure piping and so laid that no risks will arise from stumbling over it or any damage to it.

Pneumatic drills should be so designed that the piston is securely held and there is no risk of its flying out of the cylinder. It should be equipped with retaining springs, protective hoops or other devices that will reduce the risk of the tool flying out. The tool should not be shot out with compressed air but instead taken out by hand. Before the tool is changed or the appliance is used for any purpose except the ordinary one the stop valve in the compressed-air piping should be closed.

See also section 7, protection against illness due to vibration in work with pneumatic tools; section 10, protection against noise; and section 2.9, air compressors.

Powder-actuated tools

2.8.5 General

The powder-actuated tool is an indispensable aid in mounting fixtures on solid walls and ceilings when nails, bolts, etc., have to be driven into masonry, concrete, steel, wood or other material for attaching wiring, facings, plaster forms, etc., or fastening or joining building parts. Among powder-actuated tools a distinction can be drawn between those which shoot bolts into material by the gas pressure resulting from the firing of a charge, and those in which the gas pressure of a charge moves a piston which forces a bolt or a nail into the material. There are also similar appliances that work without cartridges, but they are naturally much less powerful. Powder-actuated tools are a species of firearm. Their use involves considerable risk of accidents if their design does not comply with the relevant safety requirements and they are not handled with due care. The greatest risk lies in the flying projectile which is like a bullet. A lesser risk is constituted by splinters from the material shot into. It hardly need be mentioned that as a rule the accidents are very serious, and many are fatal. Helpers and other persons in the firing area are in just as much danger as the man firing the tool. The piston appliance, which works in a very different way, contributes a relatively small proportion of the accidents due to powder-actuated tools.

2.8.6 Powder-actuated tools

The action in a powder-actuated tool is the same as in a pistol; common features are the source of energy, the cartridge, the barrel and the high speed of the projectile, the bullet or the bolt. When it leaves the barrel, a bolt has an energy between 20 and 60 mkg according to its size and the strength of the charge. If this energy is absorbed by the penetration of the bolt into the material it is harmless, but if on striking a surface the bolt is deflected or recoils the residual energy is so great that very serious injuries would be caused if a person were struck. Deflection may always occur for example in concrete when a bolt strikes an extra hard pebble, or reinforcing steel that is too near the surface; the bolt is then bent and recoils. It is just as dangerous however when the bolt hits something, for example a fragment of concrete or a join in a wall, that does not offer the resistance that was assumed in choosing the charge. Then the bolt may go through a wall or a floor and anyone on the other side would be endangered.

(a) **Technical safety requirements in the design of tools**

In some countries authorisation of powder-actuated tools is conditional on their passing a special test, and consequently when acquiring a tool is should always be ascertained whether it complies with the safety regulations in force. Generally speaking, the design of powder-actuated tools should satisfy the following technical safety requirements:

(1) the material and the construction should conform to recognised technical rules;

(2) the appliance should be equipped with safeguards that prevent a bolt from being shot into the air, and any accidental discharge while the appliance is being handled in any way. In no case should a shot be fired when the appliance is being pushed together with both hands;

(3) the safeguards should be effective when the appliance falls 3 m on to a concrete floor;

(4) the safeguards should ensure that a shot can be fired only when the appliance is equipped with a protective shield and the end of the barrel and the edge of the shield lie flat on the material to be shot into;

(5) the pressure on the material to be shot into should be at least 5 kg;

(6) the axis of the barrel should be less than 7° from the perpendicular to the surface to be shot into;

(7) the safeguards should be in good order;

(8) the protective shield should be of tough material that will hold securely any flying bolts or fragments of bolts or material. The distance between the outer edge of the shield and the barrel axis should be at least 5 cm. With changeable

or adjustable shields this distance may be reduced only at places where effective cover is afforded by adjacent building parts. For shooting in corners, or through or by projecting parts, special protective shields or other devices should be used that lie wholly on the surface to be shot into;

(9) the recoil of the appliance should be such that it cannot injure the operator when the most powerful charge and the heaviest projectile are fixed.

(b) Requirements concerning charges and projectiles

The charge of explosive should be of good quality and suited to the appliance. Cartridges should be marked with the following colours to indicate the strength of the charge:

green = weak charge;
yellow = medium charge;
blue = strong charge;
red = very strong charge;
white = extra strong charge;
black = maximum charge.

The bolts or nails should be of such calibre and shape that they exactly fit the barrel, and be so made that they are properly centred for the whole of their travel in the barrel. The nose of the bolt should be so shaped that it meets with the smallest possible resistance in the material to be shot into, and is firmly embedded in it.

At workplaces, powder-actuated tools and cartridges should be kept only in containers specially intended for the purpose.

(c) Requirements concerning operation and handling

Every powder-actuated tool should be accompanied by operating instructions drawn up by the manufacturer. These instructions together with safety rules should be handed to the operator by the employer. When he uses the tool the operator should strictly observe the instructions. Powder-actuated tools should only be used by reliable persons at least 18 years of age: they should be thoroughly trained and made familiar with the safety rules. They should be able to take the tool apart, clean it, put it together again, use it in different working conditions and explain how it works. The employer should attest the training and instruction in a certificate that the operator should countersign and keep with him. If owing to illness or other cause no trained person is available, other persons should not be allowed to use the tool. To prevent tools and cartridges from being used by untrained persons they should be kept under lock and key.

All tools should be assumed to be loaded and handled accordingly as firearms. That means that the barrel should always be pointed downwards, and never sideways, towards a body or in the air. Loaded tools should not be let out of the hands, even for a short time. Tools should be carefully cleaned and greased according to the manufacturer's instructions. If an irregularity occurs the tool should be unloaded and then taken to pieces. If the irregularity cannot be remedied the tool should be returned to the manufacturer for overhaul. In no case should the user repair it. Tools should not be used unnecessarily or for any purpose for which they were not intended.

Before firing, the operator should make sure that there is no one in the danger zone, or behind the target. If necessary places behind targets should be fenced off so that unauthorised persons cannot enter them. Powder-actuated tools should only be fired from a safe place. This applies particularly to firing on ladders and scaffolds. Since the possibility of eye injuries from splinters or dust cannot be excluded, operators should wear non-splinterable safety goggles. A hard hat should also be worn which should conform to special specifications because it has been found that the ordinary hard hats worn by building workers do not offer sufficient protection against projectiles.

If bolts are to be fired into specially shaped pieces, building parts in corners, etc., the special protective caps or other devices designed for the purpose should be used. In small confined spaces ear protection may be necessary.

In choosing bolts and cartridges account should be taken of the consistency and thickness of the material to be shot into. The charge should not be heavier than the work requires.

Bolts should not be shot into:

- walls or light structures that they could go right through;

- materials liable to splinter badly;

- springy objects from which they could rebound;

- the edges of surfaces;

- holes whose edges could deflect them.

A second bolt should not be fired into a place at which a bolt has recoiled or broken off, or pieces have broken off the material. If a second bolt is fired it should be at least 5 cm from the place.

Not every material is suitable for fired bolts. Material should be considered unsuitable when a bolt cannot penetrate it or would have its nose damaged by penetration.

Powder-actuated tools should not be used in rooms where there is an explosion risk.

2.8.7 Special types of powder-actuated tools

Special types here means piston appliances: in these the energy of the charge acts on a piston that holds the bolt and acts on the target material. With this arrangement the bolt has no movement of its own. Its penetration speed is 0 m/sec and so there is no rebound. The bolt is forced in by the piston which is retained in the tool. When the piston comes to a stop the bolt ceases to penetrate. Bolts that fail to penetrate usually remain in the tool. If a bolt is deflected by a pebble or reinforcing steel it falls down inert. The ricocheting found with tools firing projectiles does not occur.

For this reason with piston appliances a number of precautions can be dispensed with, as follows:

- the protective cap;
- restriction of the angle of inclination;
- the safeguards, up to those against firing in the air and discharge by dropping;
- specification of surface pressure;
- wearing a special hard hat and safety goggles;
- observance of a special danger zone;
- restrictions on use.

Accidents with piston appliances may occur when they are wrongly handled after a misfire.

Cartridges that have not ignited should only be disarmed by the method described in the manufacturer's instructions, and the tool should have an instrument for the purpose.

To ensure ignition a hammer weighing at least 1,250 g is necessary; only a heavy hammer will achieve a powerful ignition that will reduce the recoil and apply all the energy of the cartridge to propelling the projectile.

2.9 Other industrial equipment

Pressure vessels

Pressure vessels mean containers and appliances in which a pressure higher than atmospheric exists or can arise. The regulations to which they are subject do not, however, apply to steam boilers, transportable closed containers for compressed or liquefied gases, or gases dissolved under pressure, or to containers for flammable liquids.

The main pressure vessels used in building operations and covered by the following requirements include those listed below:

- air receivers of compressors;
- air receivers of concrete laying, and mortar transporting and spraying equipment;
- sand-blasting equipment;

- paint and lacquer-spraying equipment;
- water pressure tanks for water supply systems;
- pressure sprayers for disinfectants and impregnating materials.

Explosions caused by faulty construction or use are the chief accident risk in the use of these pressure vessels and they need careful supervision.

2.9.1 Construction and equipment

Technical rules

The materials, design, manufacture, equipment and installation of pressure vessels are governed by recognised technical rules which in different countries have been incorporated in regulations, directives, etc. These texts also lay down requirements concerning the welding of pressure vessels and the qualifications and training of welders. This is true more especially of the fashioning of the walls of vessels, the making of welded seams, and the forming of stays and flanges, because defective construction and incomplete welding may cause the welded parts to split open or break off.

Testing

Pressure vessels should not be taken into use before they have been subjected to an initial test by a specialist, or have been covered by a type test, for which a certificate has been issued. For certain types of pressure vessel periodic tests are prescribed. The scope and periodicity of the tests will depend on the capacity and the maximum permissible pressure of the vessel, matters for which vessels are divided into various groups.

All tests should be carried out by specialists authorised or appointed for the purpose.

Construction and installation

Pressure vessels over a certain size (product of pressure and capacity in litres over 200) should have inspection openings so that the condition of the coils and the welded seams can be observed. In the simplest case the pipe connections and stay openings serve as inspection openings. On large vessels peep holes, hand holes or head holes should be provided. Where these are not enough there should be manholes through which the vessels can be entered. These holes must be large enough to enable a man to get in and out without difficulty.

Plate

On every pressure vessel a plate should be affixed bearing the following particulars:

- manufacturer or supplier;

- factory number;
- year of manufacture;
- maximum permissible pressure in its gauge;
- capacity in l.

The plate should be so fastened to the vessel that it cannot be removed without damaging the test stamp. On vessels imported from abroad the particulars on the plate should be in the language of the user's country.

2.9.2 Precautions

Pressure gauge

For measuring the pressure inside, pressure vessels should have a suitable pressure gauge that gives a correct reading over a predetermined range, and has the maximum permissible pressure plainly indicated by a red mark. Damaged or inaccurate gauges should be immediately replaced by gauges in good order.

Safety valves

To prevent excessive pressures occurring, all pressure vessels should be equipped with a safety valve that acts when the maximum permissible pressure is exceeded by more than 10 per cent. When the vessel is tested the valve should be set by a specialist and protected against unauthorised alteration. In order to avoid too frequent actuation of the valve it is well to choose a working pressure lower than the permissible maximum.

Shut-off devices

There must be accessible shut-off devices in the piping as near as possible to the pressure vessel. Air receivers on piston pumps for liquids, such as those on mortar-spraying equipment, do not need shut-off devices if they are directly connected to the pump. Similarly such devices can be dispensed with on pressure vessels for oil and water separators, etc., in piping. Pressure vessels without their own pressure gauge that can be shut off separately, as with certain types of sandblasting equipment, should be equipped with a manually operated indicator. Pressure vessels in which liquid is precipitated should be so made tht the liquid can be effectively and safely drained off. Suitable devices are cocks or valves, which should be placed at the lowest point of the vessel.

2.9.3 Nature and scope of tests

The safety of a pressure vessel in operation needs to be verified by tests carried out by a specialist both before the vessel is first taken into use and periodically afterwards. The nature and scope of the tests will depend on the capacity of the vessel and the maximum permissible pressure. The relevant regulations in different countries are not uniform so that the following requirements are only to be considered as examples.

Pressure vessels with a maximum permissible working pressure p up to 0.5 ats. gauge and a capacity J up to 2,000 l, and also vessels with a maximum permissible working pressure p over 2,000 l and the product p x J not above 200, need not be tested.

Vessels with $p \leq 0.5$ ats. gauge and $J \geq 2,000$ should be tested by a specialist before they are first taken into use to ensure that the maximum permissible pressure of 0.5 ats. gauge cannot be exceeded.

Vessels with $p > 0.5$ ats. gauge and $p \times J > 200$ but less than 1,000 should be tested before first being taken into use and periodically afterwards at regular intervals.

The initial test consists of a construction test, a pressure test and a delivery test; the construction test covers the calculation, design and manufacture of the vessel, and the pressure test is a hydraulic test.

As a rule the construction and pressure tests are carried out by the manufacturers, and the delivery test at the place of use. If the vessel is covered by a type test certificate the delivery test at the place of use can be dispensed with.

The test prescribed for a pressure vessel when the place of use is changed need not be carried out on vessels used at building and erection sites. However, to prevent such vessels from being connected to supply systems at a higher pressure than the maximum permissible working pressure, in addition to the factory plate a second plate should be affixed giving the compressor output and the maximum permissible pressure to which the vessel may be connected. This second plate should be larger and more conspicuous than the first so that the important information on it can be more easily seen than would be possible on the comparatively small factory plate.

The employer should have the periodic tests carried out at the prescribed intervals by a specialist; they will consist of an internal test and a pressure test.

In the internal test the vessel will be examined for damage that may have been caused in the course of use, for example, rusting, cracks, and damaged welded seams. At the same time the safety devices such as pressure gauges, stop cocks and safety valves will be examined. The internal tests should be carried out every four years, and the pressure tests every eight years. The owner of the vessel is responsible for observing these intervals and for having the tests made by a specialist.

Special tests by specialists may be required, particularly for vessels not subject to supervision, when in the course of use damage has become apparent. For every pressure vessel for which periodic tests are prescribed a test register should be kept, to which should be attached the specialist's certificate concerning the initial tests and any supporting document. In this register should be entered the results of the periodic, and any special, tests. The register should be available for inspection by the competent supervisory authority at the place of use. For transportable vessels such as those used on building sites, it is useful to keep blueprints for inspection.

2.9.4 Operation

Persons entrusted with the care and operation of pressure vessels should be suitable for these purposes, reliable, trained and familiar with the operating instructions. In addition to the manufacturer's instructions the employer should give special instructions whenever the working conditions make this necessary. Such working conditions exist with mortar-spraying equipment, for example, which needs to be periodically opened for cleaning.

Before every occasion of use the operator should make sure that all cut-off and safety devices are effective and in unobstructed communication with the pressure chamber. During operation the effectiveness of the safety devices should be tested. The maximum permissible pressure prescribed for the vessel should not be exceeded. If it is the pressure should be reduced by suitable measures, such as shutting off the piping or opening the blow-off valve.

Pressure vessels should only be opened when the person responsible for the job has satisfied himself that no pressure remains. Even if the pressure gauge shows no pressure the blow-off valve should be opened before the vessel is opened.

If a dangerous condition arises during the operation of a pressure vessel, the vessel should be immediately laid off. Defects and damage affecting the vessel and its safety devices noticed by the operator in the course of operations should be reported forthwith to the manager or his representative. At the changeover of shifts the incoming operator should be notified of any irregularities noticed or any remedies applied.

Cleaning and maintenance work may only be done when a vessel has been relieved of all pressure. Even on vessels that are not subject to supervision, repairs, alterations and welding should only be carried out after the employer, by consulting a specialist, has made sure whether, and if so in what manner, such work can be done without impairing the safety of the vessel. Every explosion of a pressure vessel should be reported to the competent supervisory authority whether any person has been injured or not.

Compressors

2.9.5 General

The generic term "compressor" has been chosen to describe the appliances discussed in this section, but these appliances vary very widely in design and operation. The compressors used in the building industry include both compressors of gases, and in particular air for pneumatic tools and spray guns, and compressors of solids, such as the engine-driven rams for compacting earth, gravel beds, etc., or the vibrating cylinders, plates and tables for compacting concrete.

Consequently generally valid statements cannot be made about the accident risks of compressors: they include those already described in connection with pressure vessels, and also those caused by excessively high temperature of the compressed air or the use of unsuitable lubricating oil, which may explode when a certain temperature is reached. The risks of the rams comprise those inherent in all internal combustion engines and those associated with the movement of the ram. From the accident-prevention standpoint vibrating cylinders, plates and tables are similar to electrical installations and the main risk is that of contact voltage.

2.9.6 Air compressors

(a) Construction and equipment

Compressors should conform to recognised technical rules as regards materials, design, manufacture and equipment. The factory plate to be fitted on every compressor should give the manufacturer's name, the factory number, the year of manufacture, the air consumption in m^3/h or l/min, the final pressure in ats. gauge, the number of revolutions per minute, and the power consumption in HP for compressors up to 25 HP.

Compressors and connected parts of plant should be so built that liquid hammer cannot occur. If necessary special appliances such as separators should be provided. It should also be ensured that no undue overpressure occurs in the cooling system. Pressure gauges for working pressures over 30 ats. gauge should be so made that no one can be injured if they break. It should also be ensured that the intake of gases and vapours is free from dangerous admixtures, and this is particularly important with compressors for spray-painting installations.

Air compressors should be equipped with automatic devices that stop compression before the maximum permissible delivery pressure is exceeded.

Air-compressed cylinders should be lubricated with a suitable oil. Only enough oil should be used to give satisfactory lubrication, without any penetrating to the intermediate coolers, collectors and other parts of the installation. If air compressors are equipped with cooling-water jackets a device should be fitted that enables the water flow to be observed. At a suitable point between the compressor and the air receiver there should be an oil separator, unless the competent authority agrees that this is not necessary.

Valves in the air piping should be easily accessible for inspection and cleaning. Compressor valves should be regularly inspected and leaky ones should be repaired or changed. The open and closed positions of all valves should be clearly indicated.

(b) Operation

The operation and maintenance of compressors should be entrusted only to suitable reliable persons, who should be required to comply strictly with the operating instructions, especially those of the manufacturer. The employer should arrange that the safety devices are tested at suitable intervals to ensure that they are in good working order.

Safety vales and oil separators should be cleaned at least once a week. Only the cleaning materials expressly mentioned by the manufacturer should be used.

(c) Special recommendations

Air receivers, see section 2.9.

Pneumatic tools, see section 2.8.3.

Pneumatic conveyors, see section 4.3

Protection against illness due to vibration in work with pneumatic tools, see section 11.0.

Protection against noise, see section 10.0.

2.9.7 Concrete compactors

The vibrators used for compacting concrete transmit from 300 to 15,000 vibrations a minute according to type. There are internal and external vibrators. The power is supplied by electricity or compressed air.

(a) External vibrators

External vibrators do not transmit the vibrations directly to the concrete, but through the shuttering, which propagates them. They are mainly used for compacting concrete in walls and columns. Consequently wall and column shuttering should be built very carefully so that it can withstand the additional stresses set up by the vibration.

(b) Internal vibrators

Internal, or dip, vibrators consist of a vibrating cylinder that is connected to a hose or a flexible shaft and is dipped directly in the concrete. To avoid electrical accidents due to contact voltages, electric dip vibrators should be driven by current at a safe voltage of only 42-48 V, and special transformers will be required.

(c) Board vibrators

Large concrete surfaces can be compacted with so-called board vibrators. The vibrations are produced by the superimposed external vibrator.

2.9.8 Earth compactors

For compacting sand, gravel and other loose earth, vibrators can be used like those for compacting concrete. Cohesive earth such as is usually employed for filling in excavated workplaces can only be compacted by stamping. For this purpose stamps or rams driven by internal combustion engines are used: if large and powerful they need to be operated with great care.

Pile drivers

2.9.9 General

Pile driver means a machine which drives posts, girders, pipes, planks, etc., into the ground; the hammer is connected either to a guide or placed on the mast. Pile driving is generally rough work that places a heavy strain on both the machine and its equipment and makes great demands on the experience and reliability of the crew. The first requirements for safe working are thus a faultless machine of a suitable type erected on a secure base, and an experienced crew. Where these requirements are not met accidents must be expected, and owing to the nature of the work they are likely to be serious. There are manifold causes of accidents. The base on which the machine is erected may not be secure or the machine may not be safe when it has to work at certain inclinations. Pile drivers may also overturn if a track that has subsided is not promptly raised. If all parts of the machine are not constantly examined, bolts, nuts, rope ends, etc., may be loosened by the rough work or break and fall off. Other important accident causes are errors in operation, careless fastening of the hammer, inadvertent release of the hammer, failure to release the hammer from the guide when stopping the machine or when the hammer goes astray.

2.9.10 Construction and equipment of pile drivers

Pile drivers should be constructed in conformity with recognised technical rules. The winch mechanism should comply with the usual requirements for winches, unless the nature of the operations enables special appliances such as runaway preventers and brakes to be dispensed with. Pile drivers should have a plate giving the manufacturer's name, the factory number, the year of manufacture and the type. In addition, the manufacturer's instructions for operating, erecting, altering and dismantling the machine should be kept at the work site. These instructions should be in the language of the operators.

If counterweights that are not permanently built in are used, their weight and position should be indicated. With pile drivers that can work at an inclination, there should be a table giving the maximum permissible weight and length of the striking units at different inclinations.

Suitable measures should ensure that the mechanical equipment of pile drivers can be safely operated and serviced. These will include arrangements that enable all work to be done on the framework or the mast from a safe place. It should be possible to reach the rope pulleys at the top of the machine and workplaces on the mast by a ladder. Ropes should be secured by hoops to prevent them from springing out of the pulleys.

If pile drivers run on rails they should be so constructed that if an axle or a wheel breaks they cannot overturn. The carriage wheels should have flanges on both sides.

2.9.11 **Testing**

Pile drivers should be tested when necessary, but at least once a year, by a competent person, that is a mechanical engineer, foreman mechanic or pile-driving foreman. The results of the tests should be recorded in writing, preferably in a control register kept by the pile driver.

2.9.12 **Installation, operation and repair**

The erection, dismantling and operation of pile drivers should be under competent supervision, and the relevant instructions of the manufacturer should be strictly complied with. Tracks should be properly laid and there should be adequate arrangements at the track ends to prevent overrunning.

Only suitable reliable persons over 18 years old should be employed as drivers or machine operators on pile drivers. They should be familiar with the machinery and the operation of the installation. The driver has to see to the safe condition of the pile driver and in particular to see that the brakes of the winch mechanism are tested daily before work begins. He should report defects to the competent supervisor, and if a defect is dangerous he should immediately stop the pile driver.

While a pile driver is working its stability should be permanently ensured. If necessary, pile drivers should be additionally secured on the approach of a storm. At the close of work they should be blocked to prevent them from shifting.

When the hammer is raised care should be taken that the top pulleys are not overrun. Hammers that run on a guide should be released before a pile driver unit is started up. Loose detaching devices should be so kept that they cannot fall down. This also applies to tools required on the driver framework.

Work should not be done on the mast or the framework while the pile driver is operating, or moving or being tilted. It should be serviced or repaired only while it is idle.

If the machine operator cannot see work being done at the top of a striking unit by an assistant, for instance the placing of a hoop or a hood, or cannot properly watch the working of a free-fall hammer, a signaller should be appointed to ensure co-ordination.

To prevent inadvertent release of the release rope, when a free-fall hammer is being raised the rope should be laid on the counterweight end of the lever before the machine is started. The pawl should be protected by a lug or a hoop. The release line should not be fastened to the hood.

Hard hats should be worn for all work on or near pile drivers.

Silos

2.9.13 General

Silos are containers for solid material, powdery or not, that are emptied from the bottom, for example by opening gates or by using unloading equipment. Storerooms, granaries and bunkers from which material is taken from an open side or the top do not fall within the definition of silos. Materials that easily cake and so cause jams should not be stored in silos.

Risks at silos may be caused by unsuitable construction, inadequacy of structural parts or unstable installation of transportable silos: other risks arise from faulty operation. If the silo is an unsuitable shape the material may jam and the operator may disobey orders and enter the silo. When a man enters a silo even if only small masses fall or bridges formed by the material collapse when stepped on, serious and often fatal accidents may occur unless special precautions are taken.

If structures bearing or supporting the silo, and also its floor and walls are not strong enough for the stresses that will occur, and are not constructed in a workmanlike manner the silo may collapse or fall over. The same accident may happen with transportable silos if they are installed in a makeshift way, without their foundations being properly made to suit the ground conditions.

The fundamental precautions that should be taken in the construction, installation and operation of silos are as follows:

- silos should be so built that they can safely withstand the stresses that arise without deformation of the floor, walls or bearing or supporting structures;

- the shape of the silo should ensure the most favourable conditions of flow of the material so as to avoid jams due to adhesion or caking;

- the discharge openings or devices should be so constructed and arranged that they can be used safely;

- if work inside the silo is absolutely necessary the necessary precautions should be set ou in written instructions and handed to the competent supervisor who will be responsible for their strict observance.

2.9.14 Construction and equipment

In the planning and construction of silos account should be taken not only of the dead weight but also of the static load of the contents, the dynamic stresses in filling and emptying, the asymmetric or concentric forces set up by uneven distribution of the material, various atmospheric influences such as wind, temperature, snow and frost, and all other stresses that may occur in different circumstances. Silos, hoppers and discharge openings should be so planned and built that the material can flow regularly and can be removed without using the hands. The slope of sloping floors and hoppers must be steeper than the natural angle of repose of the material. All operators' stands must be accessible from stairs, fixed ladders or gangways; stairs and gangways should be protected by railings. If on open silos with an internal height of over 2 m these gangways lie less than 60 cm below the top edge of the silo, railings should also be provided on the inner side of the gangways. Fixed ladders should have back rests. Silos with an internal height of over 5 m that have to be entered to clear jams should be equipped down to the floor with vertical fixed ladders or columns of rungs or a conveyance. Seats of boatswain's chairs should be so made that the user can be roped in. Roof openings over closed silos should be large enough to allow persons to be hauled through them by a winch.

If the material in a silo is liable to jam, arrangements should be made for jams to be cleared safely by providing poker openings, vibrators, water jets or compressed-aid jets, for example. Silos in which the material may freeze should if necessary be provided with heating appliances. When selecting the heating system account should be taken of the nature of the material (explosion and fire risks) and also the possibility of poisoning by carbon monoxide, etc.

If the inside of the silos is lit by electricity, plug and socket connections and switches should be placed outside near the entrances. At these places conspicuous notices should prohibit unauthorised entry.

2.9.15 Operation

If jammed material cannot be made to move with the fixed jam-clearing equipment, suitable tools should be used from a safe place. Entrances should be kept locked, and the keys should be kept by the supervisor. If a person has to enter a silo on foot or in a conveyance the express authorisation of the supervisor should first be obtained and the following precautions should be taken:

- before the work begins filling and emptying should be stopped;

- notices should be posted with the wording:

    Do not open the gate.   Men working inside.
    Do not switch on.   Men working inside.

    The notices should be so secured that they cannot be removed by unauthorised persons;

- every person entering should be roped, and watched from a safe place by a strong person who is familiar with the work and who will keep the rope as taut as possible. It should be possible for the watcher to obtain help without leaving his place;

- for entering a silo on foot or in a conveyance only the fixed ladders or conveyances in the silo should be used. If dust may form in the silo respirators should be worn. For the whole duration of the work the person inside if on foot should be held by a rope or roped in if in a conveyance;

- the person in the silo should not be more than 1 m below the highest point in the material. Bridges and adhesions should be removed from above with pokers. If when the silo is entered it is found that the slope of the material is too steep it should be immediately reduced to less than 45°;

- idle equipment (for filling and emptying) should not be started up again until everyone has left the silo and the supervisor has given his permission.

If repairs have to be carried out in a silo that cannot be completely emptied, special precautions must be taken, such as the construction of working platforms or scaffolds equipped with railings and toeboards. In addition all the general precatuions described above should be taken.

2.9.16 Installation of transportable silos

Transportable silos as used mostly for cement, concrete and additives need to be carefully installed. The heavy loads occurring in these silos cannot be spread over load-distributing boards or unarmoured concrete slabs. Experience has shown that boards that are not strong enough sag under the heavy pressure of the supports, and heavy concrete slabs are broken off under the feet of the supports and destroyed.

Before the silo is delivered it should be decided where it can suitably be erected and how its foundations should be made having regard to the nature of the ground. Silos should not be erected near excavations for buildings or near places where trenches are to be dug for pipes. In no case should a transportable silo be erected without suitable foundations. As a rule the foundations should be made of reinforced concrete, and may be in the form of slabs, strips or blocks. The size of the foundations, the nature of the reinforcement and the quality of the concrete will be determined by the bearing capacity of the ground and the load imposed by the silo. If the type and construction of the foundations are not indicated in the manufacturer's instructions, appropriate calculations will have to be made. Care should be taken that the foundations will not freeze in winter, and that the ground cannot be softened by water or be washed away under the foundations. Erection of the silo should not begin until the foundations have properly set. The supports should be immediately anchored in the foundations because empty silos are liable to be overturned by wind.

Heating appliances

2.9.17 General

When work at building operations has to be carried on in winter weather suitable heating appliances are needed for various purposes:

- to thaw or warm building materials;

- to clear snow and ice from shuttering and reinforcement before concreting;

- to allow the chemical process of setting in concrete and plaster to take place at the requisite temperature;
- to dry buildings or rooms;
- to counteract the physiological influences of snow and cold on human working capacity by heating workplaces, workrooms and workers' accommodation and providing hot water for washing or preparing hot drinks.

The types of heating appliance used on building sites are just as various as the needs for heat. They include steam generators, "swing flame" heaters, oil-fired air heaters, heating and drying ovens for buildings, air heaters using liquefied petroleum gas, and infra-red radiators. The names of these appliances clearly suggest the risks they entail. They include:

- the risk of excessive steam pressure in steam generation;
- the general risk of fire when open fires are used;
- special risks of fire when flammable liquids are stored and used for oil-fired air heaters;
- the risk of dangerous gases, both liquefied petroleum gases for air heaters, and carbon monoxide produced in incomplete combustion.

In addition, with every type of heating appliance there are naturally the risks that persons will be burned by coming into contact with hot parts or flames, and that the fire in the appliance will spread.

The same risk occurs with excessive radiant heat when flammable materials are stored in a room where the heating appliance is installed.

Thus, whenever heating appliances are to be used on a building site, it must be carefully considered what, if any, risks will arise from conditions on the spot and the nature of the appliances, so that effective measures can be taken to counteract them.

2.9.18 Requirements for the various heating systems

(a) Steam generating plant

If low-pressure or high-pressure steam boilers are used at building sites as steam generators it must be ascertained whether and to what extent they comply with the boiler construction regulations that different countries issue. It should be noticed that some of these regulations provide for an authorisation procedure in special cases, which means that some types of steam-generating plant have to undergo an official construction test.

In general it should be required that as regards materials and construction the parts of a steam boiler installation should be so made and equipped that in the proposed working conditions they can safely withstand the estimated stresses, and will not expose persons attending or operating them or passing near them to any unnecessary risks. Consequently an installation should be equipped with instruments that will show whether the boiler contains enough water for its safe operation, and what the pressure is inside. These instruments - the water level gauge and the pressure gauge - should be in the operator's field of vision, be easily accessible from his stand, and be well illuminated. Another safeguard is a safety valve that will blow when the maximum permissible pressure is exceeded.

If boilers or their pipes are seen to be damaged they should immediately be taken out of use. To avoid burns, pipes of steam generators should be insulated in traffic or working areas. The cocks of piping should be so made that it can easily be seen whether they are open or closed. For this reason the bore of the cone should correspond with the direction of the handle. Steam-generating plant should be installed as near as possible to the place of use in order to avoid a complicated installation.

(b) "Swing-flame" appliances

For generating heat at building sites "swing-flame" appliances can be used in several ways. Without any additional equipment the burner can thaw frozen building materials or warm materials, or foundation floors or walls that are to be concreted or built on. In these appliances a mixture of fuel and air is burned. For fuel, light heating oil or diesel oil is used. If warm water has to be prepared for concreting additional equipment is required in the form of a boiler in which the burner acts as an immersion heater.

Since an open flame issues from the burner, any fire prevention regulations applicable should be observed in their use. In other respects also handling of the appliances requires great care on the part of the users if they are to avoid burning themselves and persons nearby. Special care is needed when fuelling the appliance if petrol is used. Large quantities of petrol vapour should not be allowed to escape into rooms, nor should petrol be spilled as it could be ignited by the flame of the burner. In all cases heating appliances should be allowed to cool down before they are refueled.

(c) Heating and drying appliances for buildings fired by liquefied petroleum gases (air heaters and infra-red radiators)

The gas used for these appliances is propane or butane. These gases, which are stored and transported in specially tested and authorised containers in liquid form under pressure, are a greater fire risk than benzene and are heavier than air. Unlike coal gas or carbon monoxide they are not poisonous, but there is danger to life when escaping propane has displaced so much oxygen in the air, or the gases have consumed so much oxygen in burning that not enough remains in the atmosphere to support life. Fatal cases of poisoning by carbon monoxide may occur when combustion is incomplete, which may happen when the air supply is insufficient. These risks are most likely if the appliances are installed in closed, underground or poorly ventilated rooms, and consequently their use in such rooms should be prohibited.

With all heating, drying and radiating installations the rules for the storage and use of liquefied petroleum gases as given in other sections should be strictly observed as regards both the construction of the cylinders, fittings and consuming appliances, and the storage of the cylinders; special importance attaches to the placing of the cylinders and to measures to prevent them from falling over. In the rooms to be dried the only propane cylinders should be those connected to a burner in use: spare cylinders should not be stored in the rooms. Lighted burners should be at least 2 m from any propane cylinder.

Rooms in which heating or drying appliances burning liquefied petroleum gases are installed should have openings leading to the open air and ensuring adequate ventilation or oxygen supply. In general the use of such appliances in closed rooms should be prohibited.

In excavations, shafts, trenches and other places below ground level, additional precautions should be taken when working with liquefied petroleum gases. These gases should only be used in such places if they are well ventilated. All persons engaged on the work should have been thoroughly instructed in the relevant regulations and working rules.

(d) Oil-fired heating and drying appliances for buildings

When heat at the building site is required only for heating or drying rooms, oil-fired air heaters (oil stoves) are increasingly used (Fig. 15); they work on the evaporation principle. Their output of heat is sometimes extraordinarily great and then they are a danger to life and health, and a major fire risk. Because of this risk the fire prevention regulations in some countries make the installation of oil-fired air heaters, even temporarily, for heating or drying buildings under construction, subject to an official authorisation.

For the use of oil-fired air heaters the following rules may be laid down.

Air heaters should only be installed in rooms in which no explosive gases, vapours or air-borne dusts can form. The heaters should either be firmly secured in position on an incombustible floor or fastened to strong structures of incombustible material such as walls or pillars. Between a heater and combustible material of any kind there should be a horizontal distance of at least 60 cm, and a vertical distance of at least 1 m. If the distance between the top of the heater and the top surface of the room is less than 1.5 m, this surface should consist of incombustible material. In any case, when installing a heater care should be taken that no fire risk can arise either from radiation or from the stove overturning. It should also be ensured that the stove cannot be accidentally knocked over by persons passing and the greatest care must be taken that the oil does not escape.

Oil stoves without chimneys should only be used in rooms with a height of at least 4 m and a cubic capacity of about 400 m^3. If the natural ventilation of a room is poor, mechanical ventilation will be needed, possibly by fans. If an oil stove has automatic air circulation provided by fans it should be ensured that, if the fan stops or ceases to work properly, the stove will automatically be

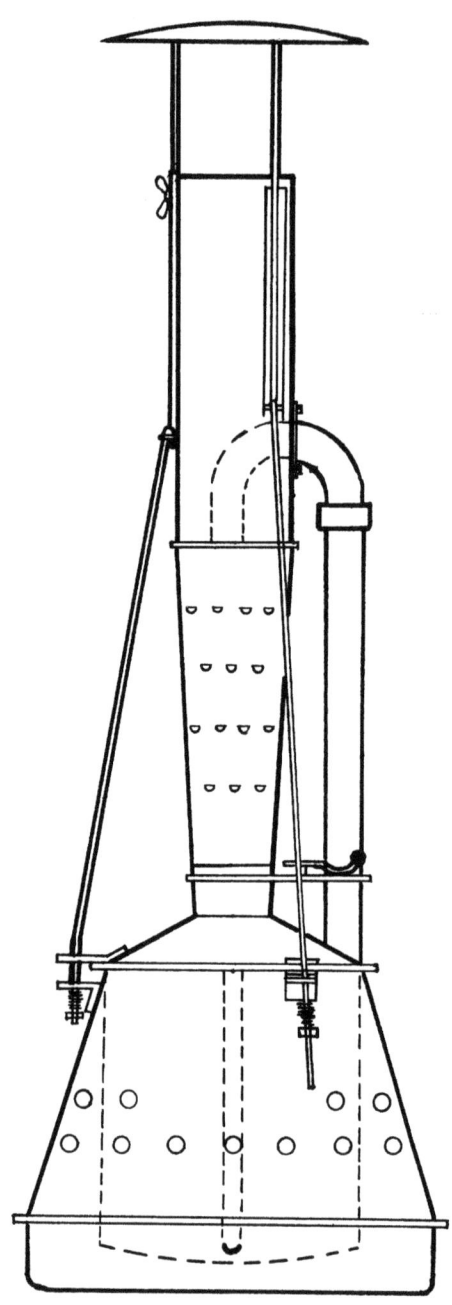

Fig. 15
Oil fired space heater

shut off. In small badly ventilated rooms a dangerous concentration of carbon monoxide may form, so the combustion gases should be led off in a chimney.

Only light mineral heating oil free of water should be used for fuel. Unrefined oils should not be used because owing to their content of sulphur compounds they can cause indispositions such as headaches. Old oil is a serious fire risk because the water that it contains will boil and cause foaming.

When a stove is used the instructions for use should be strictly observed. It should only be refuelled when the fire is out and it has completely cooled down, and the same precautions should be taken when the stove is transported. If a number of stoves are used together suitable fire extinguishers should be available. The person in charge of the stove should be familiar with the instructions for use, and carry them out. When stoves are used for drying buildings persons should be prohibited from working or staying in the rooms to be dried because of the risk of carbon monoxide poisoning.

(e) Coke-fired stoves for drying and heating buildings

Because of the formation of dangerous carbon monoxide the use of open coke fires in buildings should be forbidden, whether for heating or for drying. Coke-fired drying stoves for buildings have been developed in which complete combustion is ensured by a suitable arrangement of the air supply. If the ventilation is good there is no fear of a dangerous concentration of carbon monoxide arising.

In these stoves the carbon dioxide that forms has to be used for the chemical dissociation process and there is no need to lead off the combustion gases through a chimney or into the open. However only stoves should be used which have been certified as giving off an amount of carbon monoxide below the threshold of danger. When buildings are being dried no one should be working in the heated rooms. To ensure that this rule is strictly observed suitable precautions should be taken; for instance, at the entrance of rooms to be dried a warning notice should be displayed with wording such as: "Caution. Drying stove. No admission for unauthorised persons". At intervals of two or three hours the rooms should be thoroughly ventilated. In special cases in which adequate ventilation is not possible, as in cellars for example, flues should be provided to lead of the combustion gases. A person should only enter such rooms if a second person remains at the entrance to keep watch and to render aid if necessary.

If these stoves are only used for heating buildings, they should be connected to chimneys or flues should lead off the combustion gases into the open. When buildings are being heated the heated rooms should be ventilated from time to time. To avoid fire risks the stoves should stand on suitable bases depending on the type of floor.

3. Dangerous gases and materials

3.1 General

For accident prevention purposes dangerous gases and materials are taken to mean gases and materials whose use is attended by health, fire or explosion risks. The number of such gases and materials that may threaten building workers is much higher than is generally thought. Injuries to health may be caused by breathing toxic gases like carbon monoxide which is formed by incomplete combustion; nitrous gases like carbon monoxide which is formed by incomplete combustion; nitrous gases produced by blasting or welding; methane resulting from putrefaction in pits and sewers; and also solvent vapours. Health is also endangered by breathing siliceous dust in stoneworking or sandblasting. Again there is a risk of injury to health in working with lead paint or burning it off, for lead dust or fumes may be inhaled or may be ingested if personal hygiene is poor.

No less serious than health risks are fire and explosion risks when flammable liquids are stored and used, as fuel for heating or for machines, for example; or when explosives and gases under pressure are handled. Solvent vapours may also form explosive mixtures with air.

Burns may result from incorrect handling of hot metals, tar, asphalt, acids or alkalis.

This brief outline, which has no claim to be a complete account of the dangerous gases and materials encountered at building operations, will give an idea of the importance that must attach to combatting the risks involved.

3.2 Health risks

As indicated above, among the health risks from dangerous gases and materials distinctions can be drawn between risks from toxic gases and solvent vapours (poisoning), risks due to deficiency of the oxygen that is essential to life (asphyxiation), risks due to lead (lead poisoning) and risks due to dust (silicosis).

An aid in estimating the seriousness of the risk of harmful substances exists in the list of maximum allowable concentrations of these substances (MAC values). The maximum allowable concentration of gas, vapour or dust is defined as the highest airborne concentration at the worksite, measured at breathing level, that will not be harmful to the persons employed there if they inhale it for eight hours a day.

Toxic gases

3.2.1 Carbon monoxide

When pure, carbon monoxide is a tasteless, odourless and colourless gas, somewhat lighter than air, combustible, and when mixed with air even explosive at certain concentrations. Its properties alone clearly suggest its peculiar risks, for it is not perceptible to the senses. It is formed by the incomplete combustion of carbon or carbonaceous substances as when the supply of air or oxygen is deficient. Among the causes of incomplete combustion are premature closing of stove dampers or defective stoves. Also when internal combustion engines are left running in garages, deep shafts, or other poorly ventilated confined spaces, carbon monoxide forms in the exhaust gases. Symptoms of carbon monoxide poisoning are headache, giddiness, nausea, vomiting, general weakness, numbness and finally loss of consciousness. The gas hampers the carriage of oxygen in the blood and so leads to internal asphyxiation. In high concentrations it is instantly fatal. The mechanism of carbon monoxide poisoning is the conversion of oxyhaemoglobin to carboxyhaemoglobin which nullifies respiration. The toxicity of carbon monoxide to man may be graded as follows:

% volume CO	Toxicity
0.01 - 0.02	Tolerable for five hours without any perceptible effects
0.04 - 0.05	Respirable for one hour with hardly any perceptible effects
0.06 - 0.07	Slight symptoms after one hour
0.1	Slight symptoms after half an hour
0.15 - 0.2	Dangerous after one hour
0.2	Risk of death after one hour
Over 0.4	Fatal in under one hour

Thus with all combustion processes indoors (heating, drying, welding, running of internal combustion engines, etc.) care should be taken that the premises are adequately ventilated or the combustion gases are led off into the open. Where neither is possible coal stoves and internal combustion engines cannot be used, and another heating or power system should take their place. If persons have to enter rooms in which carbon monoxide is liable to form they should wear a respirator which must have a CO filter.

In cases of carbon-monoxide poisoning first aid consists in providing fresh air and if possible oxygen for inhalation. If respiration stops artificial respiration should be continued until a doctor arrives.

3.2.2 Methane (firedamp, marsh gas)

In the pure state, methane is a colourless and odourless gas, lighter than air, combustible, and explosive when mixed with air or oxygen in certain proportions. It is formed in places like pits, sewers, and shafts in rotting ground. If the concentration is heigh enough it arrests respiration by creating an oxygen deficiency. Since it must be assumed that narcotic gases have formed in places like pits, shafts, wells and sewers persons should only enter such places if all necessary precautions have been taken. This is particularly important because, when the precautions are inadequate, persons engaged in rescuing victims of poisoning are themselves exposed to the greatest dangers. Essential precautions are as follows:

- Persons may only enter any of the places in question on the order and responsibility of the manager or a representative expressly appointed for the purpose.

- The order should be given only after the person responsible has personally satisfied himself that the air is wholesome and that all precautions have been taken. It should be remembered that heavy gases and vapours lie close to the ground and that when sludge is stirred or scooped harmful gases are released or may form subsequently.

- Provision should be made for the continuous supply of fresh air. If sufficient air cannot be provided, air-supply respirators or oxygen breathing apparatus should be used. Rescue equipment and teams should be constantly available.

- A person entering the place should be roped and constantly watched by a reliable person. The watcher should be able to obtain help without leaving his stand. Only after help has arrived should he enter, roped, and if necessary equipped with oxygen or air-supply apparatus. All operations should be carried on under the supervision of the manager or his representative.

3.2.3 **Nitrous gases**

Nitrous gases are mixtures of various oxides of nitrogen and are highly toxic. Inhalation causes injury to health, and especially to the lungs. They are produced by the firing of explosives; if blasting is done in shafts, pits or other spaces difficult to ventilate, the gases reach such a concentration that persons entering the spaces before they have cleared may suffer serious injury. Nitrous gases are also produced by fusion and arc welding in confined spaces such as boilers and tanks, where they have the same injurious effects.

Persons should return to blasting points after firing only if measures have been taken to carry off the fumes. As with welding in confined spaces ample ventilation is necessary. If respiratory protection is required because of the impossibility of ventilating, a filter appliance with a mask but no valve should be used.

If a person has inhaled nitrous gases he should immediately afterwards be given oxygen (not under pressure) to inhale. He should remain perfectly quiet, and artificial respiration should not be practised. He should be covered over and kept lying down while he is carried away.

Harmful solvent vapours

When work is done with paints, coatings, lacquers, varnishes, impregnating materials, etc., certain liquids are required as solvents. If these solvents contain benzene or its homologues, methanol, carbon tetrachloride, di-, tri- or tetrachlorethylene, or carbon bisulphide among other substances they will liberate toxic vapours, which if inhaled will cause serious, and even fatal, injury to health.

These harmful solvents are particularly dangerous when they are used for work in confined spaces, or badly ventilated or closed rooms, that is any place with insufficient ventilation.

These solvents are sold under many trade names that do not reveal their composition but in some countries regulations require either the composition to be stated, or at least a warning to be given, on the containers in which the mixtures are sold, stored or transported for use.

Workplaces where the solvents are used need very good ventilation. If this is not possible respirators with suitable filters should be worn. A watcher who can render aid in an emergency should be present whenever solvents are used in confined spaces, or for insulation work on outside walls, or at underground workplaces alongside walls in foundation pits.

Vessels containing solvents should never be left open but should always be tightly closed. Decanting should only be done in the open or when respirators are worn. Spillage of liquid should be washed away with water. If absorbent material like sawdust is used it should be promptly removed. Windows and doors leading to the open should be opened.

Since solvents can enter the body through pores in the intact skin they should not be used for washing the hands. They are also combustible, and their vapours are explosive when mixed with air. Consequently persons handling them should not smoke or have open lights with them. See also section 3.3.

Brief accounts are given below of the properties and harmful effects of the commonest solvents.

3.2.4 Benzol

Colourless aromatic liquid. Evaporates at room temperature, combustible, and vapours explosive when mixed with air. Vapours are narcotic, and fatal if inhaled for a considerable time in closed vessels or confined spaces without ventilation. First symptoms of chronic benzol poisoning are headaches, giddiness, anaemia, tendency of mucous membranes to bleed. If respiration stops practise artificial respiration until a doctor arrives.

3.2.5 Methanol

Light volatile liquid, combustible and vapour explosive when mixed with air. The vapour irritates the eyes and the respiratory tract. Higher concentrations cause headaches, numbness, visual disturbances. If swallowed causes nausea, vomiting, colics. May also be quickly absorbed into the body through the skin. Swallowing large quantities leads to blindness. Absorption through the stomach and intestines may be fatal. If respiration stops practise artificial respiration until a doctor arrives.

When methanol is used ventilation is necessary indoors, and possibly filter respirators.

3.2.6 **Carbon tetrachloride**

Colourless, incombustible, sweet-smelling liquid, very toxic, often fatal. Absorbed through the respiratory tract and the skin. Causes headaches, nausea, vomiting, numbness, and finally loss of consciousness. Acute or chronic poisoning causes severe liver and kidney damage.

If respiration stops practise artificial respiration until a doctor arrives. Treat by rest, warmth and oxygen. If swallowed, give purge, coffee, animal charcoal but no alcohol.

If used indoors ventilation is necessary; also exhaustion of vapours at source, use of impermeable protective clothing, protective gloves, safety goggles, filter respirators.

3.2.7 **Di-, tri- or tetrachlorethylene**

Colourless liquid smelling like chloroform, incombustible. Decomposes slowly in light and air, forms hydrochloric acid and phosgene. In contact with finely divided light metal, powder or dust, may decompose and suddenly ignite and explode.

Damages the skin in the same way as all degreasing solvents. Narcotic. The narcotic effect sometimes leads to addiction, associated with visual disturbances.

If respiration stops practise artificial respiration until a doctor arrives. Treat by oxygen, rest, warmth. If swallowed and victim not unconscious, cause vomiting. Give animal charcoal, but no alcohol.

If used indoors, ventilation, exhaust vapours at source, impermeable protective clothing, gloves, safety goggles, filter respirators.

Harmful dusts

3.2.8 **Lead and substances containing lead**

See section 6.

3.2.9 **Dust containing quartz**

See section 8.

3.3 **Fire and explosion risks**

Outline of risks

In addition to the fire risks due to faulty installation of heating appliances or careless handling of flame as in soldering, welding, and cutting, there are fire risks in the storage and use of flammable liquids and gases, and these risks are always accompanied by an explosion risk. In most countries the storage and use of flammable liquids and also transportable containers for compressed and liquefied gases and gases dissolved under pressure are subject to special regulations.

3.3.1 Flammable liquids

Flammable liquids, which in addition to petrol, benzol and petroleum, include lacquers and solvents have evaporation points varying with atmospheric pressures but well below the boiling point of water. The liquids are less dangerous than their vapours when these are mixed with air or oxygen. When ignited the mixture may explode. An empty petrol drum for example is considerably more dangerous than a full one. The seriousness of the danger depends on the amount of vapour, the flash point, the ignition point and the explosive range.

The flash point is the temperature at which the liquid at normal pressure gives off so much vapour that a flame not only ignites but continues to burn. The ignition temperature is the lowest temperature at which the rate of combustion of the vapour-air or gas-air mixture is so high that the amount of heat generated is sufficient to propagate ignition. Several liquids have ignition points below 500°C and some below 200°C. Hot steam pipes may suffice to cause an explosion. There are tables of technical data concerning flammable vapours and gases from which the critical values can be taken.

Experience has shown that dangerous vapours of flammable liquids by reason of their specific gravity can easily spread over large areas, for example in underground premises where they ignite at furnaces. Air currents also carry them along without sufficient dilution so that sources of heat a long way away may be dangerous. When flammable liquids are heated, large quantitites of dangerous vapours are formed very quickly and consequently it is prohibited to heat containers of substances containing flammable solvents, for instance to make them ready for use as a coating.

3.3.2 Combustible gases

Some substances are liquid at ordinary temperatures and high pressures but gaseous at ordinary pressure, for instance carbon dioxide and butane. They are stored and transported in cylinders under pressure. If the temperature rises the liquid will need more space, but if enough space for safety is not available the cylinder may burst with explosive violence. Consequently cylinders for liquids should be protected against heat such as that resulting from prolonged exposure to the sun's rays; they should not be kept near sources of heat. They should also be prevented from falling over. Only the number required for the work in hand should be kept at the place of use.

Steel cylinders containing different kinds of gases should not be stored together, and no gas cylinders should be stored with highly combustible materials.

3.3.3 Other precautions

Fire and explosion risks with lacquers for spray painting, see section 1.9.1.

Fire and explosion risks with acetylene generators, see section 1.9.7.

Fire and explosion risks in welding and cutting, see section 1.9.5.

Fire and explosion risks when entering containers in pits, see section 1.9.12.

Requirements concerning the storage of combustible materials and liquids

Highly combustible materials and liquids should only be stored in safe storeplaces not used as living accommodation. The requirements to be laid down concerning the position and construction of premises for the storage use, etc., of combustible materials will depend on the quantities and properties of the materials and the nature of the work for which they are used. In some countries storage of these materials needs special authorisation. Supplies for immediate or daily use should be kept in unbreakable, tightly closed vessels or containers.

Corrosive burns, heat burns

3.3.4 Handling acids and alkalis

Although acids and alkalis are not extensively used in building operations we should not overlook the great risks involved in handling them. Acids and alkalis have a strongly corrosive effect on the skin and the mucous membranes, and are specially dangerous to the eyes. Splashes of acids and alkalis and also their fumes, cause swelling of connective membranes, corrosion of the cornea, and destruction of tissue. Swallowing results in severe corrosion of the oesophagus and the lining of the stomach.

Hence acids and alkalis should not be kept in vessels that could be mistaken for drinking vessels. Storage vessels should be suitably marked. In handling acids and alkalis care should be taken that they are not shaken and allowed to splash. If they have to be decanted from large vessels such as demijohns and carboys suitable devices such as tipping devices should be used. Persons handling acids and alkalis should in all cases wear safety goggles, and, if necessary owing to the quantity and concentration of the substances, gloves, aprons and rubber boots.

3.3.5 Heating asphalt and tar

Asphalt and tar are highly combustible substances, and hence should only be heated in vessels with a tight-fitting lid. While they are being heated care should be taken that they do not overflow. Vessels containing hot asphalt or tar should not be filled to the brim, or carried close to the chest or over the head. Water should never be used to extinguish burning asphalt or tar. Dry sand or suitable extinguishers should be kept in readiness for this purpose.

Tar is also injurious to health. It damages the skin and persons handling it should thoroughly clean the skin afterwards. Tar fumes irritate the eyes and the respiratory tract, and cause numbness and nausea.

3.3.6 **Melting metals**

When metals are melted, as is necessary in the building industry for preparing lead fillings, for example, the molten metal should not be allowed to come into contact with wet surfaces, and it should not be cooled by pouring water on it. Persons melting metal should wear safety goggles.

4. Personal protective equipment

4.1 General

By personal protective equipment is meant all protective equipment that is necessary to neutralise those risks of accident or ill health that cannot be completely eliminated by collective protective measures. Personal protective equipment includes hard hats, goggles, safety boots, gloves, protective clothing and respirators. To these should be added safey belts and so-called safety harness, but as they are used in the same way as catch nets as protection against falls of persons they will be dealt with in a separate section together with nets.

4.2 Hard hats

General

Statistics of building accidents show that a large number of head injuries are suffered by building workers. Most are caused by falling objects such as tools, building materials and parts of buildings and scaffolds. Because of the nature of building operations head injuries are also caused by striking the head against building parts. Experience has shown that all these accidents are more or less serious, so that one of the most important prerequisites for the safety of building workers is suitable head protection.

Wearing hard hats

Hard hats should be worn in all work and other activities in which persons might be endangered by falling objects. These will include demolition operations; erection and dismantling of shuttering; erection of scaffolds; erection of building parts; work in pits, trenches, shafts and tunnels; work on, under and near hoists, lifting appliances, cranes and transport equipment; carrying, loading and stacking goods; blasting and working with powder-actuated tools. Indeed generally speaking a hard hat should be worn for all building work, and every building worker should have one, and also every supervisor.

Specifications for building workers' hard hats

Hard hats should be tested and duly authorised, and accordingly should bear a mark testifying to their authorisation by a recognised testing institution.

A hard hat should be made of electrically non-conductive synthetic material, have adjustable ventilation, and be fitted with a face screen. A gutter round the edge will serve to drain off rainwater. The inside fittings, which should ensure that the hat sits firmly on the head, should be adjustable to fit heads in the size range 52 to 61 so that a large stock of hats of different sizes will not be needed. These fittings must be easy to clean so that the hat can be kept in a perfectly hygienic condition.

Light coloured hats are to be preferred as the wearer can be easily seen even in poor visibility, and also heat will be reflected.

Specifications for hard hats for bolt-gun operators

Hard hats for operators of bolt guns or powder-activated tools should protect them against injuries from flying bolts, parts of bolts or fragments of material. Consequently they should be made of specially strong material, difficult to penetrate or break, and should have a rim.

Testing hard hats

When hard hats are tested their shock-absorbing capacity is measured in a drop-test appliance with the kinetic energy at 4.5 kgf/m. Hard hats are also tested for resistance to penetration and breakage, lateral rigidity, resistance to abrasion and electrical insulation.

Function of the hard hat

If a hard hat is struck by a falling object, the vertical component of the energy is transmitted through the shell to the bands and fittings inside, whereby some of the energy is absorbed and the remainder is transmitted to the head. The inner surface of the shell should normally be at least 30 mm from the skull and should not move more than 2 mm nearer. This shock-absorbing distance should always be maintained.

4.3 Eye protection

General

The human eye as the most important directional organ is always turned straight towards the work and hence the source of danger and is particularly exposed to injury. The complex structure of the eye makes it understandable that even a small injury can mean a large loss of sight. However, the much greater risk of losing the sight altogether as a result of eye injuries lays a very clear emphasis on the importance of providing suitable protection for the eyes during work.

Means of protecting the eyes should be used when the eyes may be affected in any of the following ways in any operation:

- by small particles of low kinetic energy and by dust, e.g. when using hand chisels, percussion drills, pneumatic drills, etc., for cutting or breaking stone;

- by splinters and particles with higher kinetic energy, and by larger flying fragments, e.g. when grinding, cutting or sawing with fast-rotating machines, using bolt guns or sandblasting;

- by harmful visible radiation, ultra-violet radiation and infra-red radiation (radiant heat and dazzle), e.g. when welding and cutting with gas or electricity);

- by splashes of hot material, e.g. when slaking lime, handling hot tar, bitumen asphalt, etc.;

- by corrosive solids and liquids, e.g. when slaking lime, laying on mortar, handling acids and lyes, etc.;

- by irritant gases and vapours, e.g. when handling paints and insulating materials containing solvents, or working in high dust concentrations.

For all these activities and risks there are suitable means of eye protection whose construction, shape and types of glass will afford the best possible protection.

Types of eye protection

Eye protectors may be classified as follows:

- protective goggles with a separate eyepiece for each eye, with or without side guards;

- protective goggles with a common eyepiece for the two eyes;

- protective goggles with radiation filters (welders' goggles);

- goggles for protection against corrosive substances;

- goggles with an airtight rubber frame (so-called gas goggles);

- wire mesh masks (stone hewers' protective goggles);

- protective goggles fastened to head bands, caps or hard hats and worn in front of the face;

- protective screens either held in the hand or worn on the head;

- transparent covers fixed in position before danger spots or fitted as traps for shavings and splinters.

Because of this wide choice, in acquiring eye protectors, the employer or his representative should have a clear idea of the eye risks to be guarded against and of which protector would be best for the purpose.

Requirements for eye protectors

4.3.1 General

The following general requirements may be laid down for goggle frames: the total field of vision of the unprotected eye should not be reduced by more than one-fifth; the eyepiece should not slant outwards; the distance between the eyepiece and the eye should be as small as possible; the position of the eyepiece should be appropriate to the direction of the sight.

4.3.2 Protective goggles with a separate eyepiece for each eye, with or without side guards

Protective goggles should be as light as possible and capable of fitting different head shapes. Exposed metal parts should as far as practicable be corrosion-proof. The headband should be so made that the goggles sit perfectly even if the head is shaken violently. The side guards should be transparent enough to enable movements at the sides to be seen. Ventilation holes in the side guards should be so made that the protective effect is not impaired. Eyepieces should be free from defects that noticeably reduce vision.

4.3.3 Full-sight goggles

Full-sight goggles allow ordinary corrective spectacles to be worn under them without any trouble. The field of vision should not be much restricted by the frame, especially at the bottom and the sides. Contact surfaces should exert as little pressure as possible on the head.

4.3.4 Protective goggles with radiation filters

Protective goggles for welding operations and operations in radiant heat should have been tested, and bear the manufacturer's mark and an indication of their protection rating. The side guards should prevent eye damage without appreciably impairing side vision.

4.3.5 Goggles for protection against corrosive substances

The covering at the side should prevent any liquid coming from above reaching the parts of the face to be protected. At the points of contact on the face the frame should be provided with soft elastic synthetic material or rubber that is water-repellant and free from harmful substances. Ventilation holes should be at the sides and covered by splash guards.

4.3.6 Gas goggles

The frame and the enclosure should be made of rubber or an equally gas-tight material that as far as possible is also acid and alkali-resistant. The size and shape of the enclosure should be such as to ensure a gas-tight fit on heads and noses of different shapes. The eyepieces should afford the widest possible field of vision, and be mounted gas-tight. The eyepieces should be made of plain glass at least 2.5 mm thick, with cover glass over them that can be easily changed. The head bands should be at least 20 mm wide and adjustable.

4.3.7 Wire mesh masks

The wire mesh should be made of material that is mechanically strong and protected against corrosion. The colour should be dark, matt and durable; greens are preferable. The frame should be water-repellant and sewn with a strong yarn (perlon). At places in contact with the face the frame should be covered with a well and durably fastened pad of soft elastic synthetic material or rubber that contains no harmful ingredients, is water-repellant, and has a non-porous surface. The head bands should be at least 15 mm wide, long enough, adjustable, and fastened to the frame, not to the body of the mask. When ready for use a wire mesh mask should be basket-shaped and fit round the face. When not in use it should be possible by some simple means, for example by pressing a button, to collapse the mask so that it can be comfortably carried in the pocket.

4.3.8 Protective screens

All parts of protective screens should be at least fire-resistant and free from harmful materials. Eyepieces of transparent material should be free from defects that would noticeably impair vision. The eyepieces which are only intended to protect the eye from mechanical damage should transmit at least 89 per cent of the incident light. They should not create any noticeable optical distortion.

Welders' protective screens should be equipped with standard radiation filters.

Protective screens of wire mesh should be corrosion-proof. If such screens are intended to afford protection against mechanical action the mesh should be equivalent to that of wire mesh masks. The fastenings on the head should ensure a pressure-free fit.

4.3.9 Transparent coverings

Transparent coverings should be adapted to working conditions but owing to their diversity no general requirements can be formulated.

Acquisition, issue and keeping

Only eye protectors that meet the relevant requirements should be acquired. Every person in an undertaking whose work regularly exposes his eyes to danger should have a pair of safety goggles suitable for his work and for his use alone. It is specially important that safety goggles should be adjusted to fit the wearer, and that the needs of wearers of corrective lenses should be taken into account. According to the nature of the undertaking the goggles used regularly by the workers should be kept either at the workplace or in a special container that is protected against damage and dirt. If dangerous work is done only rarely eye protectors in different sizes may be kept for occasional issue.

Care and maintenance

The cover glasses used with some kinds of goggles, gas goggles for instance, should be treated in accordance with the maker's instructions. In the course of time cover glasses become cloudy and so need changing. Coating of goggles without cover glasses can be prevented by rubbing the eyepieces lightly with cleansers such as glycerine soap and then polishing them with a clean cloth. The efficacy of eye protectors should be tested at regular intervals and damaged parts should be replaced.

4.4 Safety boots

General

The building worker is much exposed to foot injuries. This is largely caused by working on insecure ground, dealing with dumped or stacked materials, handling heavy building parts and appliances, and the continually changing conditions and risks on the worksite. About 10 per cent of building accidents cause foot injuries, of which some four-fifths are punctures by nails and one-fifth toe injuries. The frequency of these injuries is evidence of the common bad habit of wearing old wornout boots or shoes in the belief that they are good enough for the building site. Even slight foot injuries are a great hindrance and make a man's working capacity open to doubt. It is absolutely necessary for building workers to wear safety boots that protect the most endangered parts of the foot during work.

Requirements applying to safety boots for building workers

If a safety boot is to prevent foot injuries during work it must afford protection against falling objects, overturning pieces of work, rolling wheels, nails, shavings, glass splinters, wire ends, etc., and also against chemicals, water and damp. Protection against the harmful effects of lime, cement and damp entails the use of specially strong and resistant material.

4.4.1 Protection against falling objects, etc.

Protection against falling objects is provided by effective toe protectors in the form of toecaps built into the boot. Under test toecaps should withstand energy of 20 kgfm developed by a hammer with a cutting edge falling 1 m, while ensuring sufficient room for the toes with a margin of 15 mm without breaking or cracking. By sufficient room for the toes is meant the space under the arch of the cap where the toes are; during the test the toes should not be touched at all. Usually steel toecaps preserve toes from injury, or at any rate from crushing.

4.4.2 Protection against penetration of nails

As protection against the penetration of nails, etc., safety boots should have impenetrable soles satisfying the following four requirements. They should be impenetrable, pliable, corrosion-proof and protect all the sole of the foot. An impenetrable sole is one that is not penetrated by a 2 1/2 inch nail loaded with 120 kg. By a pliable sole is meant an impenetrable sole that withstands 10 million flexions of the foot without breaking or cracking. An impenetrable sole is adequately protected against corrosion when it retains its protective properties after nine months' hard wear at a building site. Subject to slight differences it should correspond to the welt of the boot, and hence afford adequate protection to the whole of the sole of the foot.

Boots with so-called impenetrable inserted soles cannot be accepted as safety boots, because loose inserted soles can be removed unnoticed and the boot loses all its protective value. Impenetrable soles and toecaps should be so built into safety boots that the boot must be destroyed before the protective devices can be removed.

4.4.3 Other protection

Good results have been given by safety boots with rubber soles firmly attached to the uppers by vulcanising or welding. Rubber soles have the advantage of providing insulation against contact voltages. This advantage remains even if strips of steel are built into the sole because the foot is insulated from it by non-conducting and waterproof materials.

4.4.4 Rubber boots

Toecaps and impenetrable soles are also used with rubber boots, which afford sure protection against destructive and corrosive substances, and also against wet during work in water, mud, etc.

4.5 Protective gloves

General

The protection of hands during work is primarily a technical task for the makers of machines and appliances who should make them so that hands cannot be caught or trapped, but other means of protection are needed when the risk lies in handling sharp-edged or pointed objects that cause puncture wounds, or in loading, carrying or stacking objects that injure by crushing, or again in other operations with building materials or harmful substances. A suitable means of protection is the safety glove which every worker should be required to wear whenever possible.

Requirements applying to protective gloves

Not every glove will protect the hand during work. The diversity of jobs and the consequent diversity of functions of the hand make different types of glove necessary. The elements of manual functions are grasping, feeling, guiding, holding, turning and touching. Methods of grasping comprise the finger-tip grip, the palm grip, and the so-called key grip. With the finger-tip grip small objects are grasped by the ball of the thumb and the index finger. In the palm grip the clenched hand is reinforced by the thumb. In the key grip the ball of the thumb lies along the outside of the index finger.

According to the nature of the work and the functions of the hand that it entails, mittens, three-finger gloves or five-finger gloves can be worn.

For work requiring a sensitive touch five-fingered gloves will be needed. All protective gloves should be made of suitable material; leather has been found particularly suitable.

4.6 Respiratory protective equipment

General

In several kinds of work the production of harmful dust, mist or gas cannot be avoided. They include the following:

- quarrying and processing rocks containing quartz (sandstones, granite, silicates, etc.);

- sandblasting;

- jobs on furnace brickwork;

- burning off paints containing lead;

- spraying or dipping with the use of harmful substances or solvents.

However, dangerous concentrations of harmful gases can arise in other ways, for example, in combustion processes with an insufficient oxygen supply, as for example when buildings are being dried, internal combustion engines are left running in badly ventilated rooms, or blasting is done in pits, tunnels or shafts.

If suitable precautions are not taken in such conditions the result will be severe injury to health, if not loss of life.

In all cases in which the harmful dusts, mists or gases cannot be removed by exhaust, ventilation or other technical means, the workers exposed to these contaminants should wear respiratory protective equipment.

Respiratory protective equipment

There are filter appliances and fresh-air appliances.

With filter appliances air taken from the immediate surroundings is breathed after being cleaned; the air should contain at least 15 per cent oxygen. A filter appliance does not restrict the user's freedom of movement. Appliances for filtering dust do not afford protection against harmful gases and vapours.

With air-line respirators air is breathed after being sucked or led through a hose. The user is not dependent on the composition of the surrounding air and so the air-line respirator gives full protection. Its efficacy is not limited in time but the user's freedom of movement is hampered by the hose. In addition to filter and air-line respirators, container appliances (compressed-air respirators) and regeneration appliances (oxygen apparatus) are used to give protection against harmful gases and dusts.

4.6.1 Filter appliances

Essentially a filter appliance consists of a filter and a mask. There are appliances for protection against coarse dust and appliances for protection against fine dust. There are three grades of protection:

- grade I: protection against coarse dust;
- grade II: protection against fine dust;
- grade III: protection against airborne solids, gases and vapours.

The mask makes an airtight connection between the filter and the user's respiratory organs. When only the respiratory organs are to be protected it may be a half mask enclosing only the mouth and nose. If a specially tight and secure fit on the face is necessary, as with highly toxic airborne solids or vapours, a whole mask that covers all the face including the eyes must be worn.

Filters in protection grade I (coarse dust) are for dusts of substances that are harmless but a nuisance. Respirators for coarse dust do not protect against harmful dusts, fumes and mists of any kind.

Protection grade II can be divided into three sub-grades:

- protection grade IIa: fine-dust respirators for protection against harmful mineral dusts;

- protection grade IIb: fine-dust respirators for protection against toxic dusts, metal fumes and mists;

- protection grade IIc: fine-dust respirators for protection against radioactive dusts.

Respirators in grade II only protect against non-volatile airborne contaminants, that is those whose vapour pressure is so low that at ordinary temperatures no dangerous concentrations can occur.

Filter grade III comprises all filter appliances that retain both airborne solids and harmful gases and vapours. They have the following uses:

- against dusts and mists of volatile harmful substances which can form dangerous concentrations at room temperatures;

- in spray-painting with volatile solvents, e.g. with nitrolacquers or synthetic-resin lacquers and the like;

- against dusts and mists of harmful substances that after retention in the dust filter may give off harmful gases by decomposing;

- when there is simultaneous generation of harmful airborne solids and gases or vapours.

The filters of filter appliances may be differentiated according to the protection grade for which they are intended.

In grade I the filter consists of cloth, wadding or natural, synthetic or rubber sponge. Most coarse dust is retained, but hardly any fine dust.

Grade II filters are mostly made of prepared filter papers or felt. To prevent premature clogging by coarse dust, a coarse-dust pre-filter is often fitted.

In grade III filters a grade II filter for airborne solids is combined with a gas filter. Their useful life depends on the amount of material to be filtered out of the air.

4.6.2 <u>Fresh-air respirators (air-line respirators)</u>

Air-line respirators consist of half mask, a whole mask or a protective helmet connected to a hose through which clean, dust-free air (not oxygen) is either carried to the user or aspired by him. In the latter case half masks and protective helmets are unsuitable.

They can be used whenever the hose does not hamper work. They should be used when dangerous dusts, fumes or mists are produced in such quantities that a filter would soon be clogged. When compressed air is used an oil separator filled with activated carbon and a filter for air contaminants should be inserted in the line to remove all traces of oil from the air.

Air-line respirators should also be used when an oxygen deficiency occurs or is suspected, or when in addition to dangerous dusts, harmful gases or vapours are present in fairly high concentrations, as in containers or other confined spaces. The air-line appliance should be worn with an exhalation valve, and tight-hose connections.

Use, care and cleaning of respiratory protective equipment

Respiratory protective equipment should be used only if it has been found suitable by a competent institution. Every person exposed to dust should be provided with a well-fitting respirator. When a half mask is fitted care should be taken that the edge rests on the bridge of the nose for otherwise breathing will be hampered. Initial discomfort when wearing a mask should not lead to relinquishing its use.

Spectacle wearers who use a whole mask should wear spectacles with thin or flat head bands or side pieces.

Respiratory protective equipment should always be kept in perfect condition and carefully handled, properly stored, regularly cleaned and tested, and competently repaired. If a large number of respirators are used it will be useful to entrust their care to a special attendant. For cleaning masks, sponge filters and face sponges and lukewarm soapy water with a disinfectant should be used.

5. Safety harness - catch nets

5.1 Safety harness

General

The term safety harness embraces all personal protective equipment designed to ensure the safety of persons who are in danger of falling during their work. It includes safety belts with lifelines, retaining belts, safety ropes, protective appliances and rope unwinders.

The use of a safety harness implies that no better protection is prescribed or possible for the work to be done. See also section 1.1.8 on precautions against falls of persons.

Like any other protective device, safety harness only serves its purpose when it is made in conformity with the regulations and is properly used. Accidents occur in the use of safety harness either becuase the harness does not satisfy, or no longer satisfies, the requirements laid down for it, or because it is not properly fastened or used. However, accidents occur more frequently because people do not bother to use the harness than because it does not satisfy the requirements.

Requirements respecting manufacture

For all work that involves a risk of falls of persons and for which the use of safety harness is prescribed, only recognised tested harness conforming to generally accepted technical rules should be used. This means that the harness should have durably marked on it the manufacturer's name and mark, the year of manufacture and the approval mark. The relevant technical rules are incorporated in standards in some countries and in directives in others. Harnesses are required to have a minimum breaking strength of 1,600 kgf, which means that a body weighing 75 kg can fall 3 m and still be safely held; falls should not deform snap hooks or fittings on the harness. The structure to which the harness is fastened should also have an adequate breaking strength. In practice, it is rarely possible to test the breaking strength of the fastening point and as an additional precaution ropes with a high stretch factor should be used to absorb the energy generated by the fall.

5.1.1 Safety belt with lifeline

The most commonly used safety harness is the safety belt with a lifeline (Fig. 16). It is suitable when protection is needed against a fall from a small height. Safety belts with lifelines consist of a waist belt with one or two fastening rings and a lifeline or belt. One end of the lifeline is fastened in a fastening ring on the belt and the free end is joined to a second fastening ring by a snap hook.

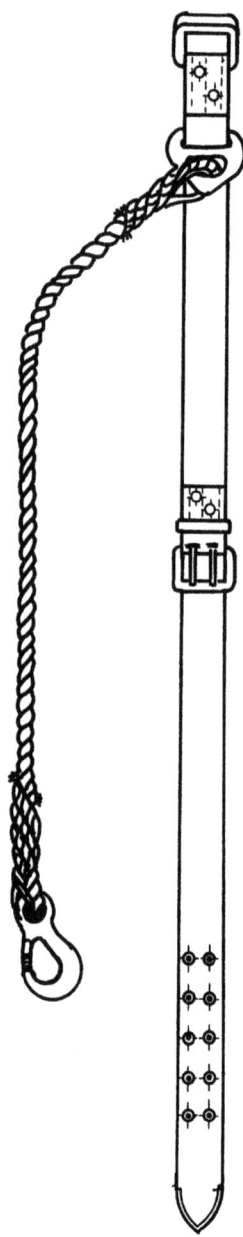

Fig. 16
Safety belt with life-line

The waist belt may be a hemp belt with leather strips or a perlon belt. The fastening device is a buckle. Hemp belts should be long enough to allow a length of at least 100 mm to be drawn through the buckle when it is adjusted for the maximum body circumference. The part beyond the buckle should have five hole positions for the buckle pins. A perlon belt should be long enough to allow a length of at least 250 mm to be drawn through the buckle when it is adjusted for the maximum body circumference, and there should also be five hole positions for the buckle pins.

The hem in which the buckle is fastened to the belt should be at least 100 mm wide and the seam should be handsewn. All fittings should be made of suitable steel or malleable aluminium alloy, and be corrosion-proof. Fastening rings and snap hooks, which should be shaped like load hooks, should be drop-forged.

Hemp lifelines should have a diameter of at least 16 mm and be made of the best long-fibre hemp. Perlon lifelines should have a diameter of at least 12 mm. Lines should not be more than 1.40 m long. Belts used as lifelines should be made of the best chrome leather without cracks and be at least 36 mm wide and 6 mm thick.

5.1.2 Safety ropes

Safety ropes are hemp or perlon ropes of medium length to which persons to be protected against falls can be fastened, indirectly by means of a safety belt, or as at quarries, mine tips, etc., directly by tying them round their bodies. The ropes are secured to a point above the workplace. Like lifelines, hemp safety ropes should also be made of the best long-fibre hemp and have a diameter of at least 16 mm. Perlon safety ropes should have a diameter of at least 12 mm.

5.1.3 Retaining belts

In contrast to safety belts, retaining belts are used to protect persons from falling from great heights and ensuring that they are not injured by being held. Consequently they should be so designed that when a person falls the sudden tensile stress caused by retention is transmitted mostly to the thigh or the seat; as far as possible stress on the abdomen should be avoided.

The width of the belt should be such that even after a person has hung for some time the flow of blood is not excessively reduced. The fastening rings required for attaching the rope should lie on the back above the body's centre of gravity. Particular care should be taken tht the fittings and connections to the rope cannot strike the head in the event of a fall. The belt should be so made that it can be put on without the help of a second person and every belt should be accompanied by instructions for putting it on properly. All bearing parts of the belt should be swaged. At a suitable place on the belt there should be an identification plate. Belts should be tested as to their behaviour under stress and be provided with an approval mark.

5.1.4 Protective appliances

Appliances for protection against falls from heights are fastened above the workplace and allow the persons roped to them to alter their distance slowly from the fastening point while the rope remains taut so tht no loop forms. If the speed of the person roped exceeds a certain limit, if he falls for example, after a short drop the rope is slowly brought to a standstill.

The most important features of an appliance are a rope drum with a rope, a device for automatically winding up the rope and a brake. The handiness of the appliance depends on its weight, which varies with the length of the rope but should not exceed 20 kg.

Since the brake of an appliance only acts when a certain speed of movement is attained these protective appliances are only suitable for work at which a free fall is possible. The appliances should only be used in association with safety belts. The fastening point of an appliance should be as nearly vertical as possible over the workplace of the protected person. Care should be taken that the connections used for securing the appliance to the fastening point have the same strength as that required for the appliance.

Appliances should be made with due regard to the conditions in which they have to be used so that their strength and working capacity cannot be impaired by corrosion, dirt or wear.

5.1.5 Rope unwinders

Rope unwinders are used for emergency descents from elevated workplaces. They enable a rope to be lowered at a suitable speed. Lowering is automatic and the person being lowered cannot interfere with it.

As regards construction and use what is said under 5.1.4 applies.

Use of safety harness

Safety harness should be used solely for the protection of persons and not for other purposes. The user should be instructed in its proper use. Damaged harness should not be used even if the damage appears to be trivial. It should not be exposed to any influences that might impair its protective properties, for instance corrosive and other erosive substances, and, above all, great heat and sparks.

Safety harness that has been strained by a fall should be immediately returned to the place of issue, and should not be used again until it has been examined by a competent person. Repairs should be undertaken only by a competent person who should observe the generally accepted technical rules.

Storage and care

Safety harness should be kept in premises that are dry but not too warm. They should not be placed near heating appliances. Safety belts should hang freely in storage. Safety ropes should hang freely or be loosely coiled. Damage from tools or sharp-edged objects should be avoided.

Safety harness should be examined by a competent person at suitable intervals but at least once a year to see whether it is in faultless condition. If any defect is found the harness should be withdrawn from use.

5.2 **Catch nets**

General

As a means of protection against falls of persons, catch nets are chiefly used in the building of halls, where they are stretched under the workplace, about the height of the trusses. In the building of houses and factories they are stretched vertically or at a slant as a covering for outside protective scaffolds. In the building of chimneys and towers it is usual to enclose the shuttering and work scaffolds completely, the nets being raised as the work progresses. The catch net is an effective means of protection against falls of persons in building operations, and even possesses some advantages over the protective scaffold. A fall on to a catch net is much gentler because even a net stretched taut is much more elastic than a scaffold floor. Consequently the risk of being injured in a fall is much less with a net, although because of its elasticity the height of the fall may be almost double that with a scaffold. With a net stretched horizontally a drop of 6 m is acceptable. But all these statements only apply if the net is of faultless construction and is properly secured.

Requirements concerning construction and use

Catch nets can be made from hemp or synthetic fibres. Because of the liability of hemp fibres to rot synthetic fibres are preferable; however, in certain circumstances synthetic fibres are affected by light, which may impair their quality, and for this reason it is advisable to use black fibres. As regards quality of the material, size of the mesh and type of manufacture, nets should conform to the standards or official regulations in force. The net and its suspension devices, and also the structural parts to which it is to be attached, should be strong enough to arrest a load of 150 kg falling 6 ft without being damaged. The mesh width of the net should be 60 mm. Since objects smaller than this can fall through, no persons should be allowed underneath. For edge and bearing ropes it is advisable to use safety ropes. The loops for suspension should be eye-splices with thimbles, and be properly joined to the net. Loops should be not more than 2.5 m apart.

Nets should be so suspended that there is plenty of room for them to stretch and rebound; it must be reckoned that because of the big stretch of a loaded net there may be a sag of as much as 1.5 m.

The fastening points for the net should be so chosen or arranged that they can withstand the tensile stresses set up by the unloaded weight and any load with a sufficient margin of safety. The more horizontal the suspension and the tighter the net is stretched, the greater will be the tensile stresses acting horizontally on the fastening points. As far as possible nets should be so placed that they extend beyond the sides of the work area to be protected.

The tauter a net is the better will it spring when an object or a person falls on it. More especially for installing large nets (10 x 10 m), it is advisable to use rope blocks and tackle with four or six pulleys, which facilitate raising the net and are convenient for tensioning it.

Care and testing

Catch nets should be handled and stored carefully. They should also be under constant supervision, because if they are not kept suitably dry when not in use they may be weakened or damaged. To test the strength of a stretched net it is advisable to drop a weight of 100 kgf into it from the height of the workplace.

HEALTH PROTECTION

General

For the purposes of the present section, health protection means measures necessary for the prevention of those diseases classified as occupational and reckoned as equivalent to occupational accidents as regards the obligations of statutory compensation.

6. **Disease due to lead and its compounds**

6.1 **General**

Handling lead and its compounds or substances containing lead entails a risk of lead poisoning. The risk is all the more serious because it cannot be revealed by either taste or smell, and the consequences of absorbing lead are not apparent immediately but only after weeks and months, or even years. In the building industry it is painters who are most exposed to the risk. However, cases of poisoning do not occur as often in the application of paints containing lead as in welding steel structures coated with lead paint, and in removing lead paint with an abrasive wheel or burning it off with a soldering lamp.

Poisoning can only occur if lead enters the blood, usually by inhalation of vapour, dust or fume, or less frequently, if the substances containing lead are swallowed. Experience has shown that an atmospheric concentration of 1.15 mg/m³ inhaled for eight hours can be considered harmless. The usual symptoms of lead poisoning are intestinal troubles, loss of appetite and vomiting; a typical symptom is abdominal pain that may develop into violent colic. With chronic poisoning there is anaemia and characteristic changes in the blood cells. Sometimes there is a lead line giving a grey-blue discolouration of the gums.

6.2 **Protective measures**

The risk of poisoning can be eliminated:

- by replacing harmful by harmless substances;

- by technical measures;

- by using personal protective equipment;

- by personal hygiene.

It is easy to replace substances containing lead by less harmful substances free from lead for there are paints on the market in which a lead compound has been replaced by another pigment and which are just as good as lead paints.

Technical protective measures

The essential technical protective measure is the use of exhaust appliances which remove dust at its source. If this is not practicable, personal protective measures should be taken, which means that respiratory protective equipment should be used, either air-line or filter appliances, depending on the working conditions. It is extremely important that the respirator should fit tight, be properly maintained and be regularly worn. See also section 2.46 on respiratory protection.

Personal protection and hygiene

If the hands and clothes cannot be prevented from coming into contact with substances containing lead, and they become badly soiled during work, suitable working clothes and gloves should be worn. To prevent street clothes from being soiled by working clothes and a risk being created in this way, separate storage facilities should be provided for each. It should be possible for the workers to wash either at the workplace or in its immediate vicinity, and for this purpose they should be furnished with a nail brush, soap and a towel. On engagement and before starting work, workers should be instructed in the risks to health and the requisite precautions. They should be required to remove dust from the hair, take off their working clothes, and throughly wash the hands and face before each meal and at the close of work. Some countries prohibit the employment of women and young persons under 18 years old on work with a risk of lead poisoning.

6.3 Medical supervision

In order to ensure complete protection of persons who have to handle substances containing lead, they should be medically examined on engagement and at regular intervals afterwards. Pre-employment examinations make it possible to avoid the employment of persons who are medically unfit to handle substances containing lead. Periodical examinations allow symptoms of poisoning to be detected in good time and a change of job to be made so that they do not become worse.

7. Carbon monoxide poisoning

See section 3 on dangerous gases and substances, part 3.2.1 on carbon monoxide.

8. Disease due to dust

8.1 General

Lung disease due to quartz dust (silicosis) is caused by the action of siliceous dust that can penetrate to the lung. The risk increases with the intensity of the dust concentration, the proportion of particles capable of penetrating to the lungs and the content of free silica (silicon dioxide, $SiO_.$). The risk exists in the mining or quarrying, manipulating and processing of sandstone, quartzite, greywacke, siliceous earth, siliceous slate, quartzite slate, granite, porphyry, pumice and kieslguhr. Free silica is particularly dangerous in the form of quartz, tridymite and cristobalite. Material containing silicates may also be a source of danger. Consequently in the building industry the following operations entail risks:

- mining or quarrying, manipulating and processing rock;
- blasting with quartz sand;
- work on furnace masonry.

8.2 Course of the disease

In all the operations mentioned fine silicogenic dust passes down the upper respiratory tract to the lymph vessels and the lymph ganglions of the lungs. The dust absorbed causes changes in the lung tissue by forming small nodules which in the course of time increase in size until finally they join to become large knots and swellings. In this way the respiratory surface, and hence the respiratory capacity of the lungs, is progressively diminished, and this may lead to cardiac and circulatory troubles. As a rule silicosis develops slowly. It is noteworthy that it can continue to develop after exposure to dust has ceased.

8.3 Protective measures

Protective measures will vary with the nature of the work, but they fall into three general classes:

- effectively exhausting the dust at the source;
- wearing respiratory protective equipment;
- periodical medical examinations.

Technical protective measures

8.3.1 Working of crude rock

In the mining, quarrying and manipulation of rock care should be taken that the workers cannot inhale dust that could cause a dust disease of the lungs. The requisite precautions will depend on the nature of the material and the manipulating and processing conditions. If there is any doubt as to whether the material to be treated involves a dust danger an expert opinion should be obtained.

(a) **Closed workrooms for stonemason's work**

For stonemason's work - sawing, milling, grinding and the like - carried on inside rooms there should be adequate airspace. Generally speaking an airspace of 25 m^3 for every person working in the room can be considered suitable, but a minimum room height of 3.5 m is desirable. Water should be laid on or suitable dust exhaust equipment provided for cleaning the rooms and their equipment. Dust should be prevented from accumulating by structural design; this can be done, for example, by bevelling windowsills, etc., and by making floors, walls and ceilings easy to clean thoroughly. The rooms should be easy to ventilate.

For stonemasons' work, the workplaces should be provided with equipment for intercepting and precipitating the dust, or the workers should be equipped with air-line respirators and the necessary accessories. Intake points for the fresh air should be protected against the penetration of foreign bodies. The air supply should be free from dust and harmful gases, and it should be possible to warm it. If compressed air is used for air-line respirators, a separator should be inserted in the line before it joins the respirator to remove oil and water.

(b) **Workplaces in the open or in open sheds**

Workplaces in the open or in open sheds should be so placed with regard to the prevailing wind that their dust-laden air cannot endanger other workplaces. Workplaces must be a sufficient distance apart; generally speaking a 4 m spacing may be considered sufficient.

(c) **Stone-working machines**

Machines for the dry working of stone should be equipped with effective appliances for intercepting and precipitating dust. Machines for wet working should be so designed or installed that the workers cannot be endangered by the dusty water spray. Crushing, sifting and screening machines should be connected to an efficient exhaust. The exhausted air should be so led off that it cannot reach workrooms and workplaces.

Workrooms, workplaces and machines should be cleaned daily but dry sweeping should never be allowed.

8.3.2 Work on furnace masonry

In furnace building, the mechanical rigging and smoothing of the surfaces of the roughly hewn bricks and trimming them by hand are particularly dangerous operations. To eliminate them as far as possible properly shaped bricks are used. If, however, bricks have to be trimmed, hand work should as far as possible be replaced by the use of power-driven grinding machines. These machines should be equipped with an efficient exhaust. Every worker hewing, grinding or scraping stone should be provided with effective respiratory protective equipment (colloid filter respirators, air-line respirators).

A considerable dust risk occurs in the demolition of furnaces built of silicate bricks. Before such a furnace is demolished sufficient time should be allowed for it to cool. When pneumatic tools are used for demolition the escaping air should as far as possible be prevented from raising dust.

When furnaces have to be repaired in great heat arrangements should be made for relays of workers, for ventilation, for screening the heat radiation and for lightening the work in other ways. For cleaning détached bricks, cleaning and grinding machines with exhaust equipment should be used. Lastly, when loose dry material like mortar and rubble is being unloaded or trans-shipped the production of dust should be restricted as far as possible by suitable measures.

8.3.3 Sandblasting

Although sandblasting is mainly done in foundries in the metal industries, it is occasionally used to clean facades or to derust steel structures. The risk can be completely eliminated if sand is replaced by a harmless abrasive, such as steel shot, steel scrap and electro-corundum. All these substances are very hard and enable the work to be done quickly. If blasting has to be done with sand, every worker should be equipped with an air-line respirator, suitable working clothes and gauntlets, and it should be seen that they are worn during the work. The respirator with the air supply should not be removed before the working area is left.

Medical examinations

Only persons who have been found fit by a medical examination, including an X-ray examination of the lungs, should be employed on work with a dust risk. The examination should be repeated at intervals as required by the competent authority. If a periodical examination shows that continuation of the work will probably be injurious to health the worker should no longer be employed on work with which he is exposed to harmful dust. Young persons under 18 years of age should not be employed on such work, but the competent authorities may allow exceptions for limited periods for young persons over 16 years of age for necessary training purposes.

A pre-employment examination should as a rule find the following persons unfit for work with a silicosis risk:

- persons who have considerable difficulty in breathing through the nose, have chronic inflammatory swellings of the mucous membranes and sinus troubles;

- persons with considerable thoracic malformations;

- persons with chronic bronchitis and advanced emphysema and asthma or an obvious predisposition to them;

- persons with pneumoconiotic changes, apart from slight changes resulting from long years of work;

- persons with pulmonary tuberculosis;

- persons with cardiac or circulatory irregularities that overstrain the heart of result in an ineffective cardiac or circulatory function.

9. Skin diseases

9.1 General

Among occupational skin diseases and distinction can be drawn between those due to handling materials that are always injurious to the skin (including unsuitable skin cleansers) and those due to allergies associated with particular materials. The frequency of occupational skin diseases in the building industry has substantially increased in recent years and more attention should be paid to their prevention. Painters are most exposed by handling paints and lacquers but handling cement may also cause skin disease.

Skin diseases of painters

When paints and lacquers are being prepared it is not possible to prevent contact or splashes of paint on the skin, especially the hands and face. While the removal of water-soluble paints and lacquers may cause inflammation and abrasion of the skin, soiling by oil paints and lacquers is more dangerous, especially when solvents such as benzine, turpentile and nitro thinners are used for removing them. These remove paint deposits from the skin, but at the same time remove much fat, especially if cleaning is repeated at frequent intervals. The skin then becomes brittle and often cracks, so that it picks up much more dirt, which causes the ignorant to use more solvent. Abrasion of the skin may be so extensive as to cause eczema.

Skin disease due to chrome cement

Recent research has shown that the well-known skin disease called mason's eczema is due to the chrome in cement. The chrome may come from the natural impurities in the raw materials of the cement, or be rubbed off furnace linings of chrome magnesite bricks, or the chrome steel of the mills for pulverising the raw material or grinding cement clinker. Efforts are being made to produce chrome-free cement. So far success has not been achieved in the attempt to find an additive that will enable the chrome content of the cement to be reduced without at the same time impairing its binding qualities.

9.2 Prevention of occupational skin diseases

General

In the course of his work the skin of the working man is exposed to many harmful physical and chemical influences. There are a number of substances which, if in daily contact with the skin, will sooner or later injure it. The principal harmful agencies are: solvents, resins, ingredients of synthetic resins, acids, lyes, tar, pitch, lacquers, lacquer thinners, numerous chemicals, oils, fats, dusts, dirt, water and also cold and heat. To these should be added strong cleaning materials and improper cleaning methods.

Measures for the prevention of occupational skin diseases include the following:

- replacing harmful materials by less harmful or harmless ones;
- changing working methods;
- improving technical equipment;
- provision and use of suitable protective clothing;
- improvement of hygiene;
- elaboration of effective skin protection;
- abandonment of cleaning methods that injure the skin.

Technical measures

Technical measures can be applied on only a limited scale. They include fitting paint rollers with a protective metal strip that prevents the paint from dripping, the provision of suitable exhaust equipment for spray painting, and the provision and use of protective clothing. This applies to all cases in which there is a risk that the hands will come into contact with acids, lyes, other corrosive or toxic liquids, or degreasing agents. The disadvantages of protective gloves are that they blunt the sense of touch in the finger-tips and prevent the removal of sweat and heat.

Barrier substances

With gloves there is a risk of substances coming into contact with the skin over the edge of the gloves or through unnoticed cracks, and remaining in closer contact than if no gloves were worn and a better means of preventing skin diseases is to use a suitable barrier substance. Before the start of work with a material that injures the skin, the clean skin is rubbed with a barrier cream. This should prevent dirt and working materials from penetrating the skin, or liquids from removing the fat from it. Although ointments, vaseline and substances containing glycerine keep the skin healthy, barrier creams and other so-called biological agents provide additional protection. The variety of substances injurious to the skin is matched by the variety of the barrier substances, of which there are two main classes - water-soluble and water-insoluble. Barrier substances have the following advantages: they can easily be applied and removed. With the water-insoluble type protection lasts for four hours; with the water-soluble type it lasts until the skin comes into contact with water. These protective preparations keep the skin healthy, let it breathe, prevent infections, and being elastic do not hamper work or allow the protective layer to be destroyed. Moreover they do not attack working materials.

9.3   Cleansing the skin

Cleansing the skin with suitable cleansers after work can also help to prevent occupational skin diseases. The choice of the cleanser depends on the nature and extent of soiling. The cleanser should be effective without injuring the skin. Most fat solvents and strong corrosives are good cleansers but they injure the skin. Where possible warm water and a mild soap should be used for cleansing, but if the hands are very dirty the use of special cleansers cannot be avoided. After the skin has been cleansed it is highly desirable to use a skin preservative.

Only by the correct choice of materials for protecting, cleansing and caring for the skin can skin diseases be prevented. In cases of doubt a doctor should always be consulted and especially a dermatologist or an occupational health doctor.

9.4   Procedure in cases of skin disease

Every case of skin disease that is presumed to be due to employment influences should be reported to the management. The treatment of the injured skin should be entrusted to an experienced doctor whose instructions should be followed strictly. Persons with a sensitive skin should not be employed on work with substances that attack the skin before they have been medically examined as to their fitness for it.

10. Protection against noise

10.1 General

Measures against noise at building operations may be divided into those official regulations which, in some countries, seek to protect the public against nuisance and injury to health from noise and from construction machinery (among other things) and those measures to protect persons actually engaged in the operations from loss of hearing.

While intense noise generally is found to be disagreeable, when the intensity exceeds 85 dB prolonged exposure to it, as at a construction machine stand for instance, leads to slight or serious loss of hearing. If the effect is only a loss of sensation, hearing will be recovered in quiet surroundings, but there will be a permanent loss of hearing if a worker is exposed to very intense noise day after day for years on end. What is at first a temporary disability develops in the course of time to irreparable damage. In addition to damage to the ear, other damage to health may be caused. Consequently, intense noise at building sites caused by machines and appliances should be combatted by suitable measures.

10.2 Noise levels

At most building machines the noise level, measured at the operator's ear, lies between 90 and 105 dB, and sometimes higher. This means that the noise made by these machines is injurious to health. The following table gives some figures for modern equipment, the noise levels being converted to those at the reference radius of 3 m recommended by the ISO. The noise intensity L is expressed in dB.

dB = decibel = 1/10 bel

Noise intensity $L = 10 \log \frac{E}{E_o}$ dB $\frac{E}{E_o}$, where E = intensity of the noise energy current in watts/cm² and $E_o = 10^{-16}$/cm². 10^{-16} watts/cm² represents the average threshhold of audibility of an adult at a frequency of 1,000 cycles per second.

Noise levels of construction machinery

Machine	Noise level in dB (A)	Type of noise
Caterpillar grader 300 HP	105	Low frequency
Vibratory screen	100	High frequency
Vibrating roller	95	Low frequency
Plate vibrator	95	Low frequency
Universal excavator	95	Low frequency
Pneumatic hammer	95	Low and high frequency
Internal pneumatic vibrator	90	Tonal
500 l free-fall mixer with electric drive	90	Low and high frequency
Small mixer with electric drive	90	Medium frequency
Internal electric vibrator	80	Low to medium frequency
Building crane	75	Low frequency

The injurious effect of noise, however, does not depend wholly on its intensity, but partly on its frequency spectrum. Shrill noise with frequencies over 2,000 cycles is considerably more harmful than low-pitched noise. If the high frequencies are filtered out of a noise by suitable means it may not affect the hearing although the intensity remains the same. Continuous noise, again, is much more harmful than intermittent noise.

10.3 Means of reducing noise at building sites

Attempts to prevent deafness take different directions, and technical measures may be distinguished from personal measures. The purpose of technical measures is to reduce the noise level to below 85 dB. Personal measures include equipping the worker with ear protectors or with suitably prepared wadding that restricts the penetration of noise into the ear.

Technical measures

10.3.1 General

Technical measures for reducing noise may be divided into short term and long term. By short term are meant all precautions appliable in the present state of technology for reducing noise. They include reducing the noise of existing construction machinery by additions and alterations, maintaining the machinery in faultless condition, and suitably arranging the building site, a measure that will benefit the public as well as the workers.

Long-term measures consist in designing new construction machinery so that it is not noisy. This can be done by using suitable componenets and enclosing the machine in a hood, and also by using remote control to keep the operators away from the machine. It may also be possible to use quieter working methods.

10.3.2 Operating condition of construction machinery

The efficacy of measures for reducing noise will depend to a considerable extent on the good working order of construction machinery. Lubrication in good time not only reduces wear on the machine but also makes for quiet running. Damaged parts should be repaired or changed. Enclosures that serve to reduce noise should not be removed.

10.3.3 Quiet building methods

The noise of building operations can be substantially reduced if they are carefully planned. In each particular case the builder should see whether, and if so which, quiet methods of working can replace noisy methods. Examples are:

- the use of an electric motor instead of an internal combustion engine;
- the use of an automatic mixer instead of a free-fall mixer;
- the use of a vibratory rammer instead of a percussion rammer;
- the use of a quiet drill instead of a percussion driver.

10.3.4 Special precautions

(a) Internal combustion engines

The noise of internal combustion engines is an aggregate of different noises. A certain amount of combustion and mechanical noise is doubtless inevitable. The noise of the intake air and the exhaust can be reduced by dampers, and that of the fan or the cool-air blower by streamlining the apertures, etc. It is also possible to provide for enclosure of the engine in the design of construction machinery.

(b) Drives

The noise from meshing of toothed gears can be reduced by accurately shaping the teeth and lubricating with an oil bath. It is possible to reduce the noise of the outgoing stream of air by means of enclosures. Enclosing the whole drive also makes for quiet.

(c)   **Free-fall concrete mixers**

The noise from a mixer is caused by the impact of the material on the drum and by the drum drive. To prevent the propagation of solid-borne noise anti-drumming material can be fitted on the drum. The drive can be made quieter with elastic elements and synthetic materials. This also applies to the feed mechanism.

(d)   **Mobile compacting appliances**

The noise from compacting appliances is composed of noise from the motor and the compactor, and the noise propagated through the casing by solid transmission. It can be reduced by making the casing in the form of an enclosure with walls that damp and absorb noise, and also dampers. Alterations can be made with a view to reducing noise.

(e)   **Pneumatic percussive tools**

The noise of pneumatic percussive tools is due to the impact of the piston on the chisel and the escape of compressed air. The intensity of the noise can be lowered by means of sheathing or dampers.

(f)   **Circular saws**

The noise can be reduced by damping the oscillation of the blade by means of elastic felt discs pressed on it.

(g)   **Pile drivers**

Most of the noise comes from the impact of the hammer on the pile and is propagaged from the surface of the hammer.

Only a slight lessening of the noise can be obtained by enclosing the hammer. Much progress towards quieter pile drivers cannot be exected until vibratory methods of driving have been developed.

10.3.5 Personal protective measures

Where technical measures for reducing noise are inadequate, personal protection may be provided such as ear protectors which block the path of the sound.

Wadding plugs are almost entirely ineffective, with the exception of glass wadding (not to be confused with glass wool). It has unusually fine fibres, and is a packing as soft as fluff for the ear passage. As a rule the fibres are given an antiseptic treatment, they are non-irritant and are well tolerated by the skin.

11. Diseases due to vibration in work with compressed air

11.1 General

As indicated in sections 2.8.3 and 2.8.4 in connection with safety requirements, it is a common feature of all pneumatic tools that they set the body, and especially the arms which hold and support the tool, in constant vibration, which may cause diseases of the bones, joints and muscles. The intensity of the vibration depends on a number of factors, such as the weight of the tool, the force of its recoil, the number of blows a minute, the length of the stroke, the amount of the operating pressure, the nature of the material worked, and the pressure exerted by the operator.

Diseases of the elbow are first in the list of injuries to health caused by the vibration of pneumatic tools. These are followed by changes in the shoulder joint and the joint between the collar bone and the shoulder blade, and more rarely muscle and nerve diseases. It is noteworthy that some workers escape all injury even after working for a long time with pneumatic tools, while others begin to complain about troubles after a few months. As a rule serious injuries to health occur only after two or three years' continuous work with pneumatic tools.

11.2 Prevention of diseases due to vibration in work with pneumatic tools

Preventive measures include:

- employment only of persons fit for the work;

- limitation of spells of work;

- use of tools whose design and equipment as far as possible prevent injury to the operator from the recoil.

11.2.1 Selection of workers and limitation of spells of work

Work with pneumatic tools liable to endanger health because of the recoil and vibration should only be entrusted to men over 21 years of age who are physically strong and sound. Spells of work should be limited, especially with heavy pneumatic drills such as are used for breaking up road surfaces, or concrete or masonry. This is best done by relieving operators about every two hours if only because handling these heavy machines is very tiring, and consequently the quantity of work falls considerably after a time. Moreover, it may be assumed that predisposition to illness increases with fatigue.

11.2.2 Technical preventive measures

All structural changes designed to improve pneumatic tools are directed towards weakening the recoil before it affects the body, but without lowering the efficiency of the machine or adding substantially to its weight, which would make it less handy. These efforts to strike a satisfactory balance between the forces acting on a pneumatic tool are subject to certain limits, not least the strength of the operator. In the light of all this a new model has been designed which is indeed heavier but does not require the operator to use any more force to pull the drill steel out of the ground or to shift the machine.

The new model is equipped with an appliance called a Pons cylinder after its inventor. It consists of a steel cylinder on which a piston slides rigidly connected to a piston rod. The rod which projects from the cylinder has at the bottom a steel sole plate covered with rubber. The piston can move in two directions; the compressed air can act on both the top and the bottom. The cylinder is connected to the compressed-air supply for the machine by a short hose. By pressing a button the upper or the lower chamber of the cylinder can be opened as desired.

When at rest the pneumatic tool stands with both points - the drill point and the point of the piston rod of the Pons cylinder - on the ground, and can be pivoted on one of them to move into a desired position without any need to lift it. When compressed air is admitted to the lower chamber of the Pons cylinder, its piston rod rises so that the drill rests only on its tip and is ready for use. If the drill jams in the ground, the concrete or the masonry to be broken up, the operator only has to let the compressed air into the upper chamber of the cylinder and then the piston rod will descend, so that the tip of the drill steel is drawn out without the operator having to exert any extra force. The advantages of the Pons cylinder are clear: notwithstanding an increase in the total weight of a pneumatic drill, the vibration at the level of the handle is appreciably less. The operator does not have to exert so much pressure as before, a strong pull is no longer needed to draw out the tip of the drill, and force is no longer needed to shift the machine from one point of operation to another. As measurements have shown, the force required to control and handle a pneumatic drill is now small, and that means a marked lessening of the recoil and vibration transmitted to the human body.

HYGIENE, FIRST AID, WELFARE

12. Hygiene

The occupation of building sites by persons for longer or shorter periods according to the scope of the project requires sanitary measures, such as the provision of accommodation, washing facilities and toilets. In the interests of the workers' health, the same care should be taken with these facilities as with all other matters affecting the work. Even if they will be needed for only a short time, they should not be makeshift.

12.1 Workers' accommodation

There will be differences between accommodation which is only used for changing clothes and resting during breaks, and accommodation in which the workers live and sleep. While the former may be on the actual site, it is better to place the latter some distance away. If convenient, accommodation for a number of building sites can be concentrated at a suitable place.

Daytime accommodation

Rooms for daytime accommodation should be large enough to allow a surface of at least 0.75 x 0.75 m. Thus a shed 2.5 x 4.5 m will suffice for 15 workers. The furniture should include a table and a seat for each worker. In the cold season and bad weather, accommodation should be heated. If a stove is installed the fire-protection regulations should be complied with. The walls near the stove should be fire-resistant and are best made or lined with asbestos sheeting. In the accommodation there should be facilities for warming food and drink, and drying wet clothes. It should be possible to open the windows for ventilation.

Living and sleeping quarters

Accommodation in which the workers spend their leisure time and sleep should have a clear inside height of at least 2.3 m. In the living rooms there should be a surface of at least 1 m² for each worker, and in the bedrooms an air space of at least 10 m³. The floors should have a warm covering. It must be possible to open the windows for ventilation, and their area should be at least one-tenth of the floor area of the room. The outside door must open outwards and it should be possible to lock it. The entrance should be provided with a wind-break. The rooms should be heatable, and there must be facilities for drying clothes outside the living and sleeping quarters. The drying space must be accessible to all workers.

There must also be means of warming food and drink. In extensive accommodation it is advisable to have a separate kitchen with cooking equipment. Kitchens should have a supply of cold, and if possible also hot, water. In addition, there must be facilities for washing kitchenware, etc.

The living rooms should afford each occupant ample sitting room at the table. They should be lighted well enough to allow reading and writing. Every worker should have a lockable cupboard or locker large enough to accommodate clothing, personal effects and small supplies of foodstuffs.

There should not be more than six beds in a bedroom and they should not be in more than two tiers. The beds should be equipped with mattresses or straw sacks, pillows, woollen blankets and linen. The linen should be changed when necessary. If shifts are worked, each shift should have its own sleeping quarters. All rooms should be cleaned regularly and kept free from vermin.

12.2 Washing facilities

Each man should have at least a washbasin. There should be an ample supply of cold, and if possible also hot, water. If special washing facilities are provided with running water, in the form of fountains, troughs, etc., one washplace for five men is sufficient.

12.3 Toilets

In daytime accommodation at building sites there should be one toilet seat for every 20 workers and in living and sleeping quarters, one for every 15. Toilets should be at least 10 m distant from living and sleeping quarters and other accommodation. If the toilets cannot be connected to a sewage system, watertight receptables should be provided; they should be emptied at regular intervals and disinfected daily. Earth privies should be avoided because of the risk to ground water.

12.4 Other facilities

In all accommodation there should be a first-aid box whose contents must be constantly replenished. If more than 50 men sleep in the accommodation there should be a separate first-aid room where the sick can be provisionally accommodated. A sufficient number of hand fire extinguishers in good working order should also be provided, suitably distributed and kept in generally accessible places.

For first-aid boxes and rooms, see also section 13 on first aid.

For hand fire extinguishers, see section 3.3 on fire protection.

12.5 Transportable accommodation

Sheds are usually erected only at places where they can remain for some time as they stand, that is at large building sites or when there is centralised accommodation. For small and medium-sized sites where the operations do not last long, transportable accommodation provided with suitable facilities has proved useful.

13. **Measures of occupational hygiene**

Occupational hygiene covers also cleanliness at workplaces, personal service rooms and accommodation, cleanliness at mealtimes, proper storage of food and drink and the provision of wholesome drinking water.

In all this the co-operation of the workers is essential and management should do everything possible to encourage high standards.

14. First aid in occupational accidents

14.1 General

Measures for the prevention of occupational accidents and those for the administration of first aid are closely related because they both serve to protect the worker. On the efficacy of first aid depends to a decisive extent the degree to which the victim will recover and hence his chances of being re-employed. It is therefore entirely justified to consider the two classes of measure to be of equal importance.

The scope of first aid in occupational accidents includes:

- displaying a notice about first aid in the undertaking or at the building site;

- keeping suitable dressings or a first-aid box in readiness;

- keeping rescue equipment in readiness as required by any special working conditions;

- training first-aid personnel;

- providing a first-aid room at large building sites;

- calling a doctor or transporting the victim to hospital.

14.2 Instruction in first aid in accidents

General

If more than ten persons are employed at a building site or other workplace instruction in first aid should be provided in the form of a poster. It should explain the general principles of first aid and deal mainly with the treatment of wounds, fractures and dislocations, injuries caused by electricity, lightning and poisons, and also with resuscitation.

First aid is only emergency treatment and must almost always be followed by medical aid. It is most important to rescue the victim and to lay him down properly, except in the case of slight injury, when the victim himself can see about first aid. It must always be ascertained whether the nature of the injury makes it necessary to call a doctor immediately or to take the victim to a hospital.

As soon as a building site is opened it is useful to indicate on the first-aid poster the address and telephone number of the nearest doctor or the nearest hospital so that in a serious emergency no time is lost in finding them out.

Rescuing and accommodating victims

The nature of the accident may make it difficult even to rescue the victim. It must be ascertained whether the victim, or even the rescuer, will be dangered in rescue operations by electricity, toxic gases, vapours, acids or falling objects. Conditions on the spot may make it necessary to use rescue apparatus.

If the available resources are insufficient for rescue purposes a public rescue agency such as the fire brigade must be immediately alerted. As regards rescue of persons from pits or containers where there are asphyxiating or toxic gases, see section 3 on dangerous gases and materials.

The procedure for rescuing victims of accidents, etc., will depend on the nature of the injury or illness. In any case, the victim must be handled carefully. The so-called Rautek hold has been found satisfactory. It is best to lay an unconscious person on his side. First aid should be started very quickly if breathing or the heart has stopped or there has been severe loss of blood or skin.

Bleeding

There are three degrees of bleeding:

- severe bleeding from arteries endangering life;

- dangerously severe bleeding when parts of limbs have been amputated or deep wounds inflicted;

- other bleeding from superficial wounds, abrasions, or injuries caused by blows, cuts, crushing and squeezing.

A special type of bleeding is bleeding through orifices in the body such as the nose, ears and mouth.

Severe bleeding endangering life can hardly ever be stopped by tying, but the wound must be pressed by the hand so hard that bleeding stops. If an artery is injured the blood will spurt out in pulses, as for instance from the arteries of the neck, collar bone and thigh, and then pressure must be applied immediately to the artery. Bleeding from the forearm, hand, calf and foot should be stopped by bending the elbow or knee tight and applying a pad or a tourniquet. For this purpose broad bandages, rubber bands and if necessary braces, waistbands, belts and folded cloths can be used, tightened if necessary by twisting a stick. However, when a wound is bound up by a layman there is always a risk that the circulation of the blood will be stopped altogether, and the nerves and tissue damaged. Bleeding from superficial wounds can be stopped by raising the injured part and applying sterile compresses. In case of bleeding from an orifice the victim should be immediately conveyed with care to a hospital.

Treatment of injuries

14.2.1 Wounds

Apart from stopping bleeding the following points need attention in the treatment of wounds.

Wounds, even if they are dirty, should not be washed, but should immediately be covered with a dry aseptic compress ready for immediate use. If one is not available, the wound should not be covered with anything else but should be left bare until the doctor arrives. Caked blood should not be removed.

Plaster will only do for superficial wounds. When the dressing is being applied the limb should be held up stiff, especially if bleeding continues in spite of the dressing.

In the case of large or deep wounds, and all wounds near joints, especially on fingers and near the knee, a doctor should be called promptly, preferably a surgeon. This applies to even the smallest wounds if they throb or prick.

14.2.2 Eye injuries

In the event of an eye injury, both eyes, the uninjured as well as the injured, should be bound up with a prepared dressing, handkerchief or the like. If there are burns from lime, acid, ammonia, etc., the eye should be thoroughly rinsed with a large quantity of water, while the lids are kept wide apart with the thumb and index finger and the victim moves his eye from side to side and up and down. An oculist must be called as soon as possible.

14.2.3 Heat burns

Persons with burning clothes must be thrown down. The flames should be extinguished by smothering them with cloths, clothing, blankets, etc., or by rolling the victim over. Clothes that are stuck on should not be removed or torn off.

Burns should be covered only loosely with sterile dressings. If the burn is extensive it should not be covered at all. In no case should powder, oil or ointment be applied to burns. Blisters should not be pricked.

Victims should be protected against loss of heat, but coverings should not be laid directly on the dressings. They should be supported by frames or chairs, or the like.

14.2.4 Chemical burns

Burns from lyes or acids must be treated as quickly as possible. No time must be lost by taking the victim to a doctor or fetching help, but the burns should be immediately and abundantly rinsed with tap water to dilute the corrosive. Then the victim should quickly be taken to a doctor, who should be told what the corrosive was.

If an acid or a lye has been swallowed, the victim should drink water or tea and then be taken to a hospital as quickly as possible.

14.2.5 Internal injuries

In all cases of internal bleeding, from the lungs, stomach, etc., the victim should be laid on his side and not disturbed. A doctor must be called at once for none but he can help. In the case of internal injuries caused by the impact of a blunt object, as when one is struck or stepped on, and the abdomen or the shoulder is affected, the victim must be immediately taken to hospital, especially if he feels sick, retches or vomits. Nothing to eat or drink should be given. The victim should be laid on his side and transported with particular care.

Fractures

A distinction must be made between closed and open fractures. With the former, the skin over the bone is unbroken; with the latter it is broken. Among signs of visible fracture are displacement, shortening, buckling, abnormal freedom of movement, visible pieces of bone. Other signs, although not so reliable, are pain, swelling, bleeding, difficulty of movement. In all serious injuries a fracture must be suspected, and everyone should behave as though there were a fracture even if only bruises, sprains and dislocations are apparent. Usually only an X-ray examination can decide.

A limb that is broken or only thought to be broken should be kept still. In no case should an injured limb be pulled, or attempts be made to straighten or set it. If there is a wound as well as a fracture, the wound should be dressed first and then splints should be put on. The splints should be so placed that both the fractured part and the joints next to it are prevented from moving. Splints - one for an arm and up to three for a leg - must be well fastened by bandages, cloths, etc. If no suitable splints are available, the arms should be tied to the body as gently as possible with triangular cloths or bandages. A broken bone should be tied to a sound one.

The foregoing recommendations do not apply to back injuries. In these a number of persons should carefully move the victim to a flat firm support such as a board or a shutter, raising him only slightly so as to avoid the risk of injuring the spinal cord.

Accidents due to electricity or lightning

In electrical accidents the current should be immediately cut off, and with low voltage (up to 1,000 V) by switching off, pulling a plug out or removing a fuse. If the current cannot be cut off immediately, the victim should be removed from the live conductors by means of a non-conducting object (wooden) or by pulling his clothes, but the victim and metal parts or a conducting floor should not be touched simultaneously.

In accidents with high voltage (over 1,000 V) the current should be cut off by an electrician or a fellow worker who is familiar with the electrical installation. The victim should not be touched before the current is cut off. If necessary, precautions should be taken to prevent the victim falling down when the current is cut off.

If the victim is unconscious because the heart or breathing has stopped, resuscitation should be begun immediately.

The foregoing recommendations also apply to injuries from lightning.

Gas poisoning

Gases that injure the lungs can be distinguished from those that do not. In any case, victims of poisoning should first be taken into fresh air or windows should be opened. If the gas is flammable no open light should be used. If poisoning is by a gas that does not injure the lungs, for example carbon monoxide, benzine and benzol vapours and solvent vapours, clothes should be removed from the upper part of the victim's body. No drink should be given to an unconscious person. The hands and the soles of the feet should be brushed or rubbed. If breathing has stopped resuscitation should be started, if possible with oxygen apparatus.

In cases of poisoning with gases that injure the lungs, such as irritant or corrosive gases like chlorine, phosgene, nitrous gases, and sulphur dioxide, the first symptoms often do not appear until some hours afterwards. The victim should be undressed, his contaminated clothes taken away, and he should be wrapped up. He should be kept quite still lying flat on his back. He should never be allowed to walk and should be carried lying down. If he is conscious he should be given spoonfuls of hot tea or coffee. Resuscitation must not be tried.

Drowning, freezing, heat stroke, sunstroke

When rescuing a drowning person the rescuer should grasp him from behind or under the armpit or chin in the interests of his own safety. If in his panic the drowning person clutches the rescuer, the rescuer should push his hand against the other person's chin and his knee against his body, and if necessary pinch his nostrils.

After landing, tight-fitting garments should be removed. The mouth should be cleared of mud and slime with the fingers. False teeth should be removed. Then the victim should be laid on his stomach, grasped round the waist on both sides, and lifted so that the upper part of the body and head hang down and water can run out. However, time should not be wasted on this, and if breathing cannot be perceived resuscitation should be begun at once.

A person who is frozen all over should be taken into a warm room and warmed as soon as possible with warm towels, hot-water bottles or friction under constant supervision. A doctor should be called as soon as possible.

With local freezing the frozen parts of the body should be slowly thawed by moving and warming the unfrozen parts. Local warming of parts of the body injured by cold is not permissible. Here again a doctor should be summoned as soon as possible. In cases of heat stroke and sunstroke, the victim's clothes should be loosened and his boots and socks taken off. He should be laid down in a shady place. If the face is mauve the head should be raised; if it is pale the head should be lowered. He should be sprinkled with water and if he is not breathing resuscitation should be begun.

Resuscitation

Resuscitation should only be practised if breathing has stopped. No drink should be given to the victim. There are three resuscitation methods: mouth to mouth, arm lift and back pressure.

14.2.6 Mouth to mouth method

The practitioner lays the victim flat on his back and kneels at the side of his head. With one hand he presses the victim's chin upwards and pulls it forward. With the other hand he presses the forehead back and keeps it in this position. Then he breathes in normally, presses his open mouth over the victim's nose and breathes into him. After rising for a moment he repeats the action. To start with, five or six breaths are given in rapid succession, and then the practitioner breathes in and out at the ordinary rate without hurrying.

14.2.7 Arm lift method

The victim is laid flat on his back, a roll of clothing is placed under his shoulders to keep the head low and the head is turned to one side. The practitioner kneels behind the head with his face turned towards the victim, grasps both arms at the elbows and swings them slowly sideways against the head counting "one and twenty, two and twenty" (breathing in). He then grasps the arms by the elbows and swings them vertically forward to the chest which he presses downwards from the sides counting "three and twenty, four and twenty" (breathing out). If a second helper is available he should at the same time massage the heart by a kneading action repeated about 100 times a minute.

14.2.8 Back pressure method

The victim is laid down as with the arm lift method. The practitioner kneels by the victim's head which he keeps turned sideways with his knee. He moves one of the victim's arms upwards between his thighs, which restarts respiration. Injured arms should lie at the victim's side on the ground. Then the practitioner lays both hands with spread fingers on the victim's back, bends the thorax forward and downwards (breathing out), counting "one and twenty, two and twenty". Then the practitioner leans back until he is erect and lifts his hands a little from the victim's thorax (breathing in), counting "three and twenty, four and twenty". He should breathe in and out about 15 times a minute.

Every attempt at resuscitation must be continued until it is successful or the doctor is sure that the victim is dead. Sometimes success comes only after hours of effort. During the mouth to mouth and manual methods of resuscitation breathing apparatus should be provided and used if possible.

14.3 First-aid supplies, rescue apparatus

First-aid supplies

For first aid in accidents on building sites appropriate first-aid supplies should be kept in readiness. It is only on very large sites in operation for a long time that first-aid stations managed by a doctor and/or nurses can be installed and it is particularly important to provide first-aid material at other sites. The nature and quantity of this material will depend on the number of workers, and will usually be specified by the competent authorities.

The number of first-aid boxes will depend on the size of the site. They should be kept at easily accessible spots. The contents should be inspected frequently, and replenished after every occasion of use.

The minimum contents of a first-aid box for small building sites, or a kit for steel erection sites, should be:

- for deep wounds: five aseptic individual dressings of the type and size of so-called dressing packs, which should not be opened until immediately before use;

- for superficial wounds (cuts, lacerations, scratches): fifteen plaster dressings of medium size, each covered with gauze to be removed before use; instructions for use with every three dressings; for burns one small bismuth burn bandage; also a triangular dressing with instructions for use printed on, six finger stalls, and six safety pins.

Where the number and severity of the injuries found to occur make first-aid boxes with the minimum contents inadequate, larger boxes of a suitable size should be provided.

Rescue apparatus

At sites where the workers are exposed to risks of asphysication or electrical accidents, rescue equipment including resuscitation apparatus should be provided, together with a trained rescue team that can practise artificial respiration.

14.4 First-aid personnel

The persons who administer first aid in accidents in the building industry should have been properly trained. If there is no works doctor, training is best given by the Red Cross or other authorised body. The number of first-aid men will depend on the size of the site. The first-aid personnel at the site should be responsible for the proper keeping and use of the first-aid boxes.

For larger sites a special first-aid room may be necessary. A first-aid attendant should always be present there. He will keep the first-aid register, in which all cases of first aid with the name of the victim and the nature of the injury will be recorded.

15. Medical care

15.1 General

Medical care of workers serves to protect their health at the workplace, and should help to maintain and improve their health and efficiency, eliminate occupational health risks, and detect illnesses and other injuries to health in good time.

15.2 Works medical service

The works medical service is organised by the undertaking and may be established in accordance with regulations, recommendations, collective agreements or decisions of the management. It requires doctors, their assistants, suitable premises and the necessary equipment, medical and other. As a rule the service will be in the charge of a full-time works doctor. It may be assumed that a full-time works doctor can look after 2,000 to 3,000 workers. Undertakings with 500 to 2,000 workers may combine to establish a joint medical service. Undertakings with less than 500 workers may share in some other undertaking's service.

15.3 Qualifications and functions of the works doctor

15.3.1 Qualifications of the works doctor

Only a person authorised to practise medicine can be a works doctor. The works doctor, whether full time or part time, should on appointment or after a reasonable time possess the qualifications required for the practice of occupational medicine.

15.3.2 Functions of the works doctor

The functions of the works doctor include the following:

- supervision of the health of the workers in the undertaking by initial and periodical medical examinations, advice in the consulting room of the medical service, and other preventive medical measures in the undertaking;

- medical examinations required by labour, safety or other similar regulations if the doctor possesses the necessary authorisation;

- medical care and first aid in accidents and cases of acute illness; follow-up treatment in agreement with the victim's doctor and if necessary with the insurance agency concerned;

- inspection of the undertaking in company with the management, the works council, the safety engineers, the safety officials, government or other inspectors, experts, etc.;

- training of first-aid personnel in collaboration with the institutions concerned, and collaboration in the organisation of first-aid services in undertakings;

- advice in connection with the planning and installation of new plant, work study, and the development of new processes, equipment and materials;

- advice in matters of community welfare, convalescence and housing, and in other matters in which medical advice may be useful;

- collaboration in the prevention of occupational accidents and diseases, the arrangement of premises and workplaces, the regulation of working tempo, breaks and shifts, measures against noise and atmospheric contamination at workplaces, lighting, ventilation, air-conditioning, the provision of safe and convenient machines and tools, the provision of suitable protective clothing and other personal protective equipment, and in other matters of health protection at work;

- collaboration in arrangements for changing jobs on health grounds, resuming work after illness or accident, protecting young and women workers, protecting maternity, and employing elderly workers;

- collaboration in the supervision of sanitary and general hygiene measures and other arrangements to maintain health and promote social welfare, more especially the washrooms, cloakrooms, rest and meal rooms, etc., sports grounds, works kitchens and convalescent homes;

- collaboration with the workers' doctors, industrial physicians, official doctors, the labour administration, and doctors of insurance and other institutions.

THE ORGANISATION OF WORK

16. **General**

Accident prevention and health protection are a synthesis of technological, organisational and educational measures. Notwithstanding the importance of technical means of accident prevention and health protection in ensuring safe equipment, processes and places, the additional necessity of organisation and education should not be overlooked; on these largely depends the effective application of technical measures at the right time and the right place.

Although safety and health measures affect the lives of all workers in the general complex of building operations their importance is not necessarily self-evident. They should be recognised by the management and fitted into the everyday life of the undertaking, like any other measure. Accident prevention and health protection are tasks for management and it should take appropriate measures to ensure that:

(a) working equipment is designed and constructed in accordance with the technical requirements of safety, used in conformity with the regulations in force, properly installed, dismantled and attended, frequently inspected as to its safe condition and maintained in that condition;

(b) the building and other materials used satisfy quality and safety requirements, and in particular are stored, transported, handled and processed as circumstances demand;

(c) the workplaces and trafficways are laid out safely and kept safe;

(d) the safety devices prescribed for the different operations are provided and used, and any other necessary safety measures are taken;

(e) safety instructions are provided concerning the conduct of work and the behaviour of the workers;

(f) the workers are selected and trained for their jobs with due regard to their qualifications and physical fitness, instructed in the risks and the regulations concerning them, and provided with the necessary personal protective equipment;

(g) the measures required by the laws and regulations in force concerning first aid and accommodation for workers at building sites are taken; and, lastly,

(h) the undertaking's safety organisation for internal supervision is set up and the necessary administrative measures taken to ensure that it works.

17. Organisation of undertakings and building operations

17.1 General principles governing the organisation of undertakings

Every undertaking in the building industry, every building site and workplace, needs a definite form of organisation to ensure the smooth progress of work; and on this the technical and economic efficiency of the undertaking will largely depend. The type of organisation will be determined in a large measure by the type and size of the undertaking and arrangements cannot be the same everywhere; variants will be needed in specialised undertakings or unusual projects. Fresh organisational problems arise with every new project, although their range may be limited in a particular class of projects. The organisation of site very much depends on the organisation of the undertaking itself and every shortcoming in it will be reflected at the site.

Accident prevention and health protection should not be looked upon as separate tasks. All the measures comprised in the organisation of the undertaking should promote safety, whether they concern the planning or preparation of work, the provision or use of appliances, the employment of persons, reporting, management and supervision, or, last but not least, the functioning of the undertaking's safety organisation.

17.2 Planning and preparation of work

General

Planning and preparation of work are industrial functions which have long been matters of course in the stationary industries. Assembly lines in these industries are only possible because the whole sequence of operations has been precisely planned in advance in all its detail, and it has been ensured that, at the proper times, sufficient workers with suitable knowledge and qualifications will be available, that they will be provided with all the necessary machines, appliances and tools in working order, and that sufficient quantities of all necessary materials of the best quality will be ready for use.

In the building industry it was long held that such planning of operations was impracticable, and undertakings relied on improvisations, to the detriment of profitability and also of safety. It is true that in view of the variety of building operations as well as the variety of scope, purpose and situation of building sites no tried uniform scheme of organisation can be used, and often the builder lacks time because the interval between concluding the contract and beginning the work is so short. Incomplete planning data when the contract is concluded, or subsequent changes by the builder may nullify the most careful site planning. Similarly long spells of bad weather or frost can upset all calculations concerning the duration of a project.

In spite of all difficulties, knowledge gained in the static industries, suitably modified may be valuable in the management of building operations and planning and preparation take their full place in the organisation.

Purposes of work planning and preparation

Among the principal purposes of work planning and preparation is to establish plans for site equipment, and timetables for operations and lists of requirements in the way of appliances, machines, materials and personnel. In addition there are questions of power supplies, traffic, transport, storage, accommodation for workers, etc. to be settled. Other important questions concern shift, night and winter work. It is of prime importance that account should be taken of the requirements of accident prevention and health protection in all the measures decided on in the planning and preparation of work. As a rule the persons responsible for the planning and preparation of work are not the same as those who will later direct or carry it out (and who will therefore be responsible for carrying out the necessary safety and health measures); it is therefore essential that the former should be as familiar with safety and health requirements as the latter.

17.3 Building site equipment

The most important task of the planning and preparation staff at a building undertaking is the equipment of the site; this includes the provision of all machines, appliances, tools, magazines, accommodation etc., and their co-ordination with each other and the site as a whole. The smooth progress of a project will depend considerably on the suitability of equipment, the proper choice of machines and appliances and their correct use. Smooth progress includes freedom from accidents and damage and proper equipment of a site is not only economically important but will also determine the level of safety there.

Provision and use of appliances

The acquisition of any kind of industrial equipment should not be decided by economic and technical considerations alone. In building operations appliances must stand up to all sorts of stresses and meet all sorts of needs, so that it is not easy to say whether this or that type of appliance is to be preferred irrespective of cost considerations, but among the factors of decisive importance in the choice is the extent to which its design and construction satisfy safety and other related requirements (as set out in sections 1 through 5 in some detail).

Not every appliance on the market will meet these requirements and before acquiring one an undertaking should examine it in the light of the regulations in force: the manufacturer or vendor should be asked to confirm in writing that the appliance supplied complies with all safety requirements in its design and construction.

At the same time the undertaking is bound to have appliances maintained and used properly; this means that, before every occasion of use, all industrial equipment should be inspected as to its good working condition, and that if any defect is found it should be remedied in a workmanlike manner before the equipment is used again. No piece of equipment should be taken from the store to the building site before a competent person has examined it and found it fit for use. When an appliance or a machine is delivered at the site it must be accompanied by the manufacturer's instructions for maintenance and use, for its safe condition will depend largely on strict compliance with these instructions. With the primitive means available at many building sites, necessary repairs cannot be properly carried out, and makeshift repairs are frequent causes of accident.

The management has also to ensure that those appointed to use equipment are properly trained for the purpose, partly because the life of the equipment depends to no small extent on their reliability.

Transport and traffic lanes, store and processing places

The way an appliance is installed and related to the building is equally important for economy and safety. This applies with particular force to lifting and transport appliances, for every simplification and curtailment of transport operations makes for a lower accident frequency. But the safety of transport operations at a building site depends equally on the condition of the transport and traffic lanes. Transport lanes should be prepared and if necessary reinforced to suit the quantity and weight of the appliances, structures and materials to be moved on the site. No less important is the proper arrangement of the storage and processing places for building materials. Unnecessary moving of materials and difficulties in loading and unloading due to lack of space are causes of common disturbances and hence of accidents. Planning not only pays, but also promotes safety.

In planning the equipment at sites the position of the workers' accommodation is very important whether daytime accommodation for changing clothes and occupation during breaks, or living and sleeping quarters. Neither the accommodation nor the approaches to it should be endangered by the building operations and they should be outside the danger zones of machines and cranes.

Collaboration between a number of undertakings

Where a number of undertakings work side by side or in collaboration at a building site it may easily happen that friction arises between them or they may endanger one another. There should therefore be agreement on the measures to be taken by each at the time when the equipment of the site is being planned. If in special cases the equipment of one undertaking has to be used by the workers of another, even for a short time, this should be agreed in advance.

It should be made quite clear (and, if necessary, confirmed in writing) where responsiblity lies for operating the equipment; this applies equally to the common use of machines, lifting appliances and scaffolds. If common use makes alterations in installations necessary it should be decided in advance who should make them and under whose responsibility.

18. Employment of personnel

18.1 General

As already mentioned the tasks of management are not exhausted by providing safe equipment and taking the precautions for the protection of the workers in particular circumstances; it has also to regulate the employment of the personnel to appoint persons and to say how they are to work. This entails ascertaining whether each worker is fit for his job. If he is, he should be instructed in all the risks to which he will be exposed at work and the means at his disposal for protecting himself against them. Lastly, management has to put all work under competent direction and supervision, functions that include issuing the orders required for the maintenance of safety and supervising their execution.

18.2 Qualifications of workers

Principles

The suitability of workers should be examined from various standpoints. A worker should only be given a job for which he is suited by age, sex, physique, health and knowledge. This obligation to examine suitability makes heavy demands on management in the building industry, for with constantly changing jobs requirements concerning suitability, knowledge and experience also constantly change considerably. It is not enough to examine suitability only at the time of engagement; it should be repeated with every new job because the nature of the job and the project generally may impose new requirements.

Employment of young persons

When young persons are employed it is most important to observe the minimum ages for employment laid down in regulations. Some occupations prohibited for young persons and relating either to equipment or to processes are discussed in the section on technical measures. Young persons, apprentices and trainees, and especially newcomers to the undertaking, should be instructed in the special risks of the work. Older workers with whom young persons work should be instructed to advise them until they are familiar with the peculiarities of the undertaking, the equipment and the work. As a rule young persons should not be employed on dangerous work.

Employment of women

In some countries the employment of women in building operations is governed by regulations. If the regulations allow women to be employed at all in these operations, they are usually confined to jobs that do not require any great physical exertion, and so are excluded from the lifting and carrying of heavy loads. Like young persons, women should not be employed on dangerous work.

Physique and health

Assessment of suitability must be based to some extent on physique and health. That cripples should not be employed on scaffolds, working platforms, etc., that can only be reached by ladders is just as self-evident as the need for sound limbs for operating the machines and appliances used in the building industry. If workers suffer from physical defects that are not immediately apparent they should inform the management about them. This is particularly necessary when a worker suffers from giddiness, or epilepsey or the like.

If necessary in such cases a medical examination should determine whether, or to what extent, physical defects impair working capacity and constitute an additional risk for the worker concerned.

Pre-employment and periodical medical examinations

For all work that may be injurious to health, medical examinations to determine fitness are necessary. Such work includes, especially, work with dangerous gases and materials such as siliceous dust, materials containing lead, and solvents. An examination may also be necessary for work with pneumatic tools where the recoil may cause damage to health.

Everyone operating heavy construction equipment such as cranes, excavators and loaders should be medically examined to ascertain whether his eyesight and hearing are adequate, and whether he suffers from defects that make it advisable for him to follow some other occupation.

In addition to the pre-employment examination periodical examinations at regular intervals may be necessary. If the medical examination shows that a worker runs the risk of an occupational disease occurring, recurring or worsening, he should be kept out of the employment concerned until he is completely cured, or if he is found to be hypersensitive to the action of harmful materials and equipment, permanently excluded from this type of employment.

Knowledge and experience

Every occupation in the building industry, whether it is operating machines, lifting appliances, transport equipment or other equipment, or erecting structures, scaffolds and the like, requires some knowledge and experience. This applies even to simple jobs like lifting, carrying or stacking building materials, and working in excavations, that is, jobs that are usually done by unskilled workers. In order to work correctly and at the same time safely a man must possess certain skills, which must first be acquired. Hence the question whether a worker is suitable for a certain job always implies the question whether he possesses the necessary knowledge. In assessing a man's suitability for a job the management should satisfy itself that he indeed has the necessary knowledge and experience: sometimes a practical test may be necessary. For example a man who is to be employed on a circular saw should demonstrate his knowledge and not be given the job merely on his assertion that he is familiar with circular saws. To some extent proof of capacity is a guarantee of good work and conducive to proper care in the operation of machines and appliances; serious damage may be caused by faulty operation and maintenance.

A high standard should be applied when assessing the knowledge and experience of persons who are to be entrusted with the direction of a project or an operation. Not every foreman, superintendent or manager possesses the knowledge and experience required for special projects and operations, even if he has passed the appropriate vocational tests. Projects vary too widely as to type, scope and situation to allow capacity to be standardised. New materials and processes make it necessary even for persons otherwise suitably experienced to be given an opportunity of familiarising themselves with the innovations before they are made responsible for work involving them.

18.3 Training and retraining

Among the essential conditions of safe working are the training and retraining of all workers. In particular all on-site training should include instruction in the basic rules of occupational safety and the technical safety requirements applying to the workplace, job or process in question. The undertaking should therefore take advantage of every opportunity, wherever and however provided, to develop the occupational training and retraining of its workers, and in addition should take its own steps in these directions.

Since every absence of a skilled worker or a supervisor on account of illness, accident or leave, requires a replacement of equal value, it follows that all training that includes instruction in safe working is of economic importance to the undertaking.

18.4 Duties of workers

All legislation on labour protection generally, and labour protection in the building industry in particular, is based on the principle that responsibility for the application of all necessary measures rests on the employer or the management and subsidiarily on the supervisory staff. Apart from this, the example set by executives and the interest they show in matters of accident prevention and health protection are decisive in achieving safety at the site.

But independently of the role of executives, all building workers are under an obligation to promote safety, and according to their position and duties may even be required by law to do so. One can differentiate between the general safe conduct of their work and the special duties associated with the operation of equipment or in carrying out certain jobs.

General duties of workers

The workers are bound to comply with the regulations issued for their protection and the protection of their workmates, and all orders and instructions of their employer and his supervisory staff. They have also to take the precautions necessary for their own and their workmates' safety and to do nothing that might endanger themselves or others. They should use the safety devices and protective equipment provided by the undertaking. If safety devices are lacking they should ask for them in good time (without prejudice to the employer's duty to provide them). They should immediately remedy defects in equipment or report them to the employer or his representative.

Workers should not alter equipment without authority, or remove safeguards or put them out of action or damage them. Equipment and safeguards should only be used for the purpose for which they were intended.

Workers are only allowed on the workplace to the extent necessary for the performance of the work assigned to them. They may only enter and leave the workplace by the entrances and exits provided for them.

At places with a fire or explosion risk workers may not smoke or have open lights. Alcohol should not be taken during working hours. Clothes and head-dress should not be liable to endanger the wearer. Workers should not go barefoot on buildings and other workplaces.

On the way to and from work the official traffic regulations should be observed. Bicycles, motor cycles and motor cars should be in a safe condition.

Special duties of workers

Workplaces and trafficways should be kept tidy; waste and rubble and remains of building materials should be removed promptly. Projecting nails should be removed from timber etc.

(In all work only suitable good-quality building materials, appliances and tools should be used.)

When working at elevated and similar dangerous places workers should take precautions against falling. Objects should not be thrown down unless the place below is protected by watchmen or by fencing. Before anything is thrown a loud warning should be given, and the thrower should satisfy himself that the place below is clear and barricaded, or otherwise protected.

Before ladders are used they should be inspected as to their safe condition: uprights should not be botched and damaged and missing rungs should be replaced by sound runs of the same kind. Ladders should be so set up that they cannot slip, overturn or sage excessively.

Scaffolds should not be used until they have been completed. Persons should not jump on to scaffold floors. Scaffolds should not be overloaded. Makeshift scaffolds of loose stones, cases, etc. should never be built.

Machines, appliances, lifting appliances etc. should only be operated by persons who are familiar with them and have been appointed for the purpose. Maintenance and repair work and cleaning must not be done on machines and appliances in motion. Workers should not remain in the danger zone of excavators, loaders, etc.

Containers for transport should not be overloaded. Persons should not be carried on lifting appliances intended only for carrying goods. Loads should be so slung that they cannot slip out of the lifting gear.

Electrical installations should only be built, altered or repaired by electricians. Loose conductors should be carefully handled and protected from damage due to crushing, being run over, etc. When working near live conductors workers should always keep at such distance that a hand or a tool cannot come into contact with the conductors. If a safe distance cannot be maintained work should only proceed after a competent person has switched off the conductor, or covered or enclosed it.

On work where the formation of harmful gases, vapours or dusts cannot be prevented the arrangements made by the employer should be strictly followed. If necessary, respiratory protective equipment, gloves and other protective equipment should be used.

When acids, other corrosives, hot tar or asphalt, and the like are handled shaking and splashing should be avoided.

Oxygen cylinders, fittings and piping should not be allowed to come into contact with grease and oil.

Before persons enter shafts, sewers and other underground places special precautions should be taken, and the permission of the employer or his representative should be obtained.

19. Management and supervision

19.1 Appointment of responsible managers

All building operations should be carried out with great care and in conformity with generally recognised technical rules and so should be directed by a reliable competent person. If he cannot remain permanently on the site, he should be replaced during his absence by an equally competent person. When these persons are appointed it should be ensured that they possess the knowledge and experience demanded by the degree of difficulty of the work. They should be familiar with the kind of work carried on by the undertaking.

The site director, his deputy and all other supervisory staff are responsible for the application of the necessary labour protection measures equally with the employer who has legal obligations in this respect. Their duties and responsibilities follow automatically from their appointment, but it will be advisable to specify them in each case; this can conveniently be done by means of a signed contract between the parties.

All supervisors should be required to set a good example to the workers in the observance of the safety rules, and so demonstrate the great importance that the management attaches to safety. On the attitude of the management and the supervisory staff depends the success of the undertaking's efforts to prevent accidents and protect health.

Like the workers, the supervisory staff need constant instruction in the safety measures to be applied in the undertaking, and consequently they must be supplied with the relevant regulations, directives, etc. In addition they should be enabled to attend safety courses.

19.2 Establishment of supervision

As small and simple building operations which do not require expert direction a supervisor should be appointed if more than two workers are employed. Naturally, such operations should receive attention from the management even if only occasionally. In particular, before jobs are allocated the management should decide which precautions should be taken, and give the necessary orders. The supervisor should be held responsible for ensuring compliance with any orders given. The supervisor's main task is to prevent workers from taking unauthorised action and so endangering themselves and their workmates.

20. **Collaboration of the works council**

If there is a works council it is bound, like the management, to participate in the prevention of accidents and the protection of health. One of its tasks is to ensure, through local arrangements based on the law or collective agreements, that the laws, regulations, collective agreements or internal instruments protecting the workers are properly applied.

It has the general duty to concern itself with accident prevention and health protection including application of the relevant regulations, and to support all agencies in the accident-prevention field by giving encouragement, advice and information. Its duties include co-operation in all cases in which circumstances make it necessary to depart from the provisions of the regulations in force, and the management should consult it when new protective devices, processes, etc. are being introduced or tested. If the council is to perform its duties properly the management should inform it of all the accidents and cases of occupational disease occurring in the undertaking. In some countries the works council is empowered by law to countersign the accident reports the mangement is required to send to the authorities.

21. Organisation of safety

21.1 General

In principle all technical safety measures in an undertaking are part and parcel of the ordinary operations, and the promotion of safety is not to be regarded as something separate but it has been found useful in undertakings of a certain size to establish a special safety organisation without prejudice to the obligation of all supervisory staff to co-operate in safety activities. In different countries the establishment of a safety organisation may be a voluntary act of the employer, the result of agreement between employers' and workers' organisations, or a legal obligation. In contrast to the duties of managerial and supervisory staffs who are responsible over-all for accident prevention, the duties of the safety organisation lie more in the directions of watching, verifying and advising. The establishment of a works safety organisation gives practical effect to the idea of self-supervision, and shows a positive attitude towards accident prevention. It is therefore a particularly valuable contribution to the cause of safety. Experience shows that good safety organisations often do more for safety at work than regulations.

21.2 Works safety organisations and their tasks

As a rule the safety organisation has no direct power of command and consequently no punitive powers, these being exclusively reserved to the supervisory staff. Its function above all is to work for the smooth co-operation of all persons concerned with safety in the undertaking, to draw attention to shortcomings and defects, and to furnish information to the management so that it can take necessary remedial action.

The nature, composition and method of appointment of works safety organisations are not uniform in different countries.

One or more safety officers may be appointed according to the size of the undertaking: there should be one on every building site employing more than 20 workers. Where there are a number, the most experienced should be the leader. In large and complex undertakings there may be a safety engineer who will be a full-time or a part-time official according to the size and circumstances of the undertaking. The persons working for safety in the undertaking may form the safety committee, which may also include representatives of the management, and of the workers, and heads of departments concerned with questions on the agenda. If the undertaking has a safety engineer he may be chairman of the committee.

Safety officers

The safety officer should support and advise the employer in the carrying out of safety measures and act as an instructor to his workmates. In particular he has to keep constant watch to see that safety devices are in place and that the workers are working in a safe way.

This means that safety stewards should possess certain qualifications and be carefully selected. The safety steward should have been a long time in the undertaking and be well acquainted with the equipment, the work and the workers. He must know his job, and enjoy the trust of his employer and his workmates, and so be both a valuable support and a good comrade.

When a safety officer is appointed his duties should be fully explained to him personally, and his appointment should be announced by posting up a notice or calling a meeting of the workers. The occasion will remind the workers of the importance that the management attaches to safety.

Instructions should settle when and how the safety officer should make his inspections and what action he should take in particular circumstances. He should always remember that he has only advisory functions and his success will depend on the extent that he does so and acts accordingly. His activities include the following:

- learning and observing the relevant safety rules;

- exemplary conduct at work;

- verifying the presence of the necessary notices;

- ensuring good housekeeping in all workplaces;

- verifying the completeness and proper storage of the first-aid equipment (if there are no trained first-aid men);

- participation in investigations of accidents;

- noticing defects during his inspections and, depending on their nature, remedying them by his own efforts, reporting them to the competent supervisor, or reporting them in writing to the management;

- verifying the presence, condition and use of safety devices, personal protective equipment, working clothes and footwear;

- picking out important danger points and systematically inspecting them;

- instructing young and new workers;

- educational activities among the workers generally;

- verifying observance of the regulations concerning minimum age of employment;

- participation in inspection of the undertaking by outside authorities;

- participation in monthly talks on safety matters with the employer or his representative.

This large range of duties entails continued training of the safety steward by means of courses organised either by the undertaking under the safety engineer or provided by outside bodies.

The safety engineer

The safety engineer may be a part-time or a full-time official according to the size of the undertaking. He advises the management on all matters of accident prevention and health protection, and accordingly must have been suitably trained. He should understand people and industrial life, be generally well educated with professional qualifications. If he is part time it should be ensured that his work does not suffer from lack of time: he must have enough time which he can devote entirely to safety. The full-time safety engineer, usually only employed in large undertakings, is as a rule supported by a safety department with its own office facilities and in which all the safety stewards assist the safety engineer.

The chief duties of the safety engineer are as follows:

- supervision of the undertaking's equipment and operations as regards compliance with the relevant regulations and internal instructions in the field of accident prevention, the removal of dangerous conditions and the prevention of dangerous practices;

- collaboration in the planning and preparation of work to safeguard the interests of accident prevention, and in the ordering of new working equipment and personal protective equipment;

- advising and collaborating with the management in accident prevention matters;

- appointment and supervision of safety officers;

- collaboration with the works council and with the official labour inspectors;

- working out of a safety programme for the undertaking;

- inspection of sanitary installations;

- collaboration in the job training of workers;

- safety training and general safety propaganda among the workers;

- preparation and evaluation of the meetings of the safety committee of the undertaking;

- settlement of all matters requiring action for safety purposes.

The safety committee

The safety committee may consist of:

- the employer or his representative;

- the safety engineer, or if there is none, another representative of the management;

- a representative of the workers;

- the safety officers;

- other persons as needed (for example, representatives of the buying department, the personnel office, the planning office);

The regular meetings between the management, the safety stewards and the works council, prescribed by law in some countries, serve to determine future tasks, as well as to discuss matters raised by the various departments.

The safety committee serves the following purposes:

- exchange of experiences;

- discussion of accidents that have occurred and means of preventing recurrences;

- assessment of the safety situation in the undertaking since the last meeting and the conclusions to be drawn from it;

- discussion of training, education and propaganda activities in the undertaking;

- drawing up the safety programme for the next month;

- discussion of means of stimulating co-operation in safety activities (for example, proposals for rewards, determining prizes, releasing workers for training, etc.).

The works safety committee is one of the most valuable weapons in the war on accidents and diseases.

www.ingramcontent.com/pod-product-compliance
Ingram Content Group UK Ltd.
Pitfield, Milton Keynes, MK11 3LW, UK
UKHW021318180426
11947UKWH00015B/1294